The Middle Class Comeback

The Middle Class Comeback

Women, Millennials, and Technology Leading the Way

Munir Moon

MGN Press

MGN Press
640 Maple Avenue
Torrance, CA 90503
info@mgnpress.com

Printed in the United States of America

First Printing, 2016

10 9 8 7 6 5 4 3 2 1

Publisher's Cataloging-in-Publication data

Moon, Munir
The Middle Class Comeback: Women, Millennials, and Technology Leading the Way/ by Munir Moon.
p. cm.

ISBN 978-0-9913721-7-1 (e-book)
ISBN 978-0-9913721-6-4 (Hardcover)
ISBN-13: 9780991372164
ISBN-10: 0991372166
Library of Congress Control Number: 2016908136

1. The United States – Politics. 2. Current Affairs. 3. Healthcare. 4. Economy – United States.

For
Khatoon & Atif

Table of Contents

Introduction

The middle class is the hallmark of the American dream and the backbone of a free and strong society. That dream has been shattered in recent decades for many Americans. Most of the media and politicians define the middle class by income, which has been stagnant at best or declining at worst for the average American. However, it is the cost of living as well as higher taxes that are larger factors adversely impacting the middle class. Moreover, government policies have exacerbated the middle-class decline.

This book argues that the cost of education, healthcare, housing, and taxes have increased at a much higher rate than the income. One can look at the middle-class challenge as a glass half full or half empty. The premise of this book is that the glass is half full and that the middle class will come back despite all the doom and gloom about its demise.

Middle-class America is defined as a family of four including two school-aged children earning between $51,000 and $123,000 per year. They aspire to attain affordable healthcare, college education for their children, a home, and retirement security.[1] This book focuses on the jobs and ever rising cost of healthcare, education, and taxes—most of which can be attributed to the government's counterproductive policies and dysfunctional political structure. This book highlights various components that will result in the middle class comeback instead of an in-depth analysis of each category in the interest of keeping the book short. Almost 400 citations are provided for further reading.

Yes, it is true that the middle class has suffered the most in recent history while most of the economic gains have been realized by the top one-percenters. The top 0.1 percenters are worth as much as the bottom 90 percent of the U.S. households.[2] That disparity would serve as the impetus for the middle-class comeback by challenging the old way of doing business, both in private companies and government agencies.

One of the reasons for that optimism is women gaining ground with men in almost every segment of society, such as higher education, medicine, and business. Moreover, women control $11 trillion, or almost 40 percent, of the nation's investable assets.[3] Women make up half of the population, yet they are still well under-represented in society when it comes to making decisions in private and public deliberations on the economy, education, healthcare, politics, and social issues. As women gain more parity with men, such as in the job market, they will contribute more money to the family income and will lead the way toward the middle-class comeback.

Millennials, who represent 25 percent of the adult population,[4] will be the largest voting bloc over the next few decades. They are projected to inherit $30 trillion from their baby boomer parents. Millennials are transforming existing industries while creating new ones whether it is in communication, media, or goods and services. They are creating jobs, enhancing purchasing power, and improving the quality of life for the middle class.

The technology will force the existing private and public institutional structures to be efficient amid the digital age. Over the last thirty years, technology has made astounding changes. Smartphones and other technological products that are very complex yet very simple to use will continue to add productivity to the economy. Just as the auto industry had a major impact on the growth of the middle class in the early twentieth-century, technology will do the same in the twenty-first century. This will result in the transformation of education, healthcare, the political system, and the tax structure. As a result, their costs will decrease, which will increase the purchasing power of the lower and middle class.

It would be naïve to think that change will come overnight, but the transformation of healthcare and education is already underway. Moreover, the

2016 election was a watershed moment in United States history in all likelihood. The American public is so frustrated and angry with Washington and professional politicians that they are willing to accept a businessman like Donald Trump, who has no political experience, as the leader of the free world.

Rising healthcare costs have taken more and more money out of the pockets of middle-class households. They have been and will continue to be a drain on the national economy as the aging population grows under the current healthcare delivery model. The Affordable Care Act (ACA) primarily focuses on health insurance coverage for every American rather than reducing healthcare costs so that it is affordable.

The unintended consequence of the ACA is that the businesses have and will continue to shift the healthcare costs to employees by increasing deductibles and copayments to meet the mandate and maintain their profitability. As consumers are forced to pay more out-of-pocket expenses toward their healthcare costs, they will have more incentives to shop for a better value. That will bring competition to bear, as it does not exist under the current healthcare delivery system. Combining competition with technological innovations will redefine healthcare and make it more affordable.

College education is another factor that has been consuming the greater share of an average middle-class family's budget. The nation still uses an eighteenth-century Prussian model imported from Germany to educate children, a production model where students move up a grade every year. However, the advent of online learning and the skyrocketing tuition costs will make the current model to earn a bachelor's degree a relic of the past. College tuition could be reduced by as much as 50 percent by deploying online learning and reducing the physical presence of the students to two years through apprenticeships, internships, and study abroad programs.

Middle-class America has been devastated by the 2008 housing crisis where the median American family lost 39 percent of its net worth from $126,000 in 2007 to $77,300 in 2010, three-quarters of which can be attributed to the home value. The housing sector has been improving as families regain financial stability while mortgage rates are at a historic low since the financial crisis of 2008. However, the demographic shift due to millennials may redefine whether home ownership is as important as it was to the previous generation.

Americans have been taxed everywhere they look; some are direct, and others are disguised as fees, fines, penalties, or anything but the word "tax." The penalty imposed by the ACA if one does not have health insurance coverage is one of the newest. Then there is a 48-cent federal and state tax on every gallon of gas.[5] A total of $15.30 out of $100 in gross working income in Social Security and Medicare tax disguised as insurance are paid by the employee and employer combined.[6] Moreover, there are numerous fees and charges on cable, telephone, and other utility bills that can go up to 35 percent of the bill.[7] A disproportionate burden of all these taxes falls on the middle class since they represent a larger percent of their earnings compared to the top one-percenters.

The cost of healthcare, education, housing and taxes have reached to such a point that people are going to look for alternatives, and that is where the disruptive innovators come into play. They are challenging the age-old legacy model of running healthcare delivery, or a one-hundred-year-old production model educating American children. The political disrupters will challenge the existing tax structure that becomes complicated with every tax reform. One of the consequences of those reforms is that most of the recent economic gains have gone to the top one-percenters.

This book takes a look at how the middle class will come back despite some of the depressing headlines in the news about its demise.

One

The Middle Class

The lower and middle class represent the majority of Americans. This group will gain strength and influence in the coming years as it challenges the one-percenters' power in shaping the policies and affairs of the country. The 2016 election may serve as a transforming event in history where the policies affecting the middle class were in the front, middle, and back of America's public discourse.

Most Americans consider themselves to be the middle class. According to a survey by *The New York Times* in 2005, only one percent of the respondents considered themselves to be "upper class" and only seven percent considered themselves part of the "lower class." The rest, 92 percent, considered themselves middle class.[8] Most of the middle-class people are working people who do not have enough wealth to invest. However, a 2012 survey report by Pew Research states that only 49 percent (from 92 percent seven years earlier) of adults describe themselves as "middle class," which seems to be manifested during the 2016 election cycle for a change.

There is a common desire among all Americans, regardless of their socio-economic background, to have financial security and a better life for their children. They want to have affordable healthcare for their family, provide education to their children so that they can have a better future, own a home, and save for retirement. So how is the middle class defined?

There is no standard definition of the middle class. It has been defined in various ways. Some definitions are purely based in economic terms while the others are defined based on a combination of values, expectations, and aspirations. Shifting social values and expectations appear to influence whether one considers himself or herself as part of the middle class. Therefore, it is important to understand what *middle class* means through various lenses.

Economic Definition

Income is the sole measure by economic analysis to define the middle class. Some of the definitions include absolute dollar value income, income relative to median, or income level relative to the poverty line.[9] According to the White House Task Report[10], that estimate ranges from $51,000 to $123,000 for a married couple with two school-aged children in 2008. The reason for a wide range is that the cost of living in different parts of the country varies drastically where even a higher end middle-class family feels the budget squeeze.

The standard definition of household income is broken down into five quintiles, 20 percent each, the lowest, second, middle, fourth, and the highest income. According to a Congressional Budget Office (CBO) report, households with income before federal tax of $45,300 fall into second quintile, and those in the fourth quintile make $97,500 or less based on 2011 data.[11] Households between the second and fourth quintile are considered the middle class according to a report by the White House.[12]

Another definition is by the median income where 50 percent of the households make less and 50 percent more. The median income grew from $52,623 in 1990 to $57,843 in 1999 and then trended down to $53,657 by 2014 (Fig. 1.1). This reflects about a six to eight percent decline from its peak and gain from its bottom of twenty-five years ago.

A Pew Research Center study examined the changing size, demographic composition, and economic fortunes of the American middle class in 2015. It defined the middle class as Americans whose annual household income is two-thirds to double the national median, about $42,000 to $126,000 annually in 2014 (adjusting for inflation) for a household of three. Under this definition, the middle class is about 50 percent of the United States adult population in 2015, down from 61 percent in 1971.[13]

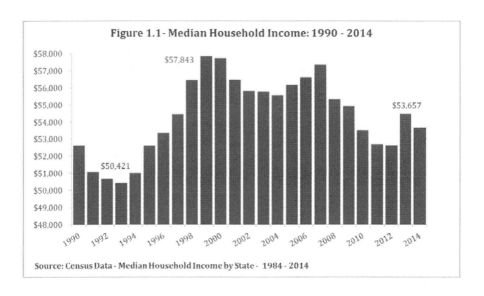

Figure 1.1- Median Household Income: 1990 - 2014

Source: Census Data - Median Household Income by State - 1984 - 2014

Self-Reported Definition

Another approach is simply asking people to identify their status in society. According to a Pew Research Survey in 2012, the median response for a family of four earning $70,000 considers themselves to be the middle class. Some estimate a median income of $85,000 in the East, $60,000 in the Midwest, and $70,000 in the South and West qualifies for the middle-class status.[14]

Qualitative Definition

By some accounts, the middle class seems to be defined by values, expectations, and aspirations rather than income. Income, however, is critical in order to realize the aspirations. The middle-class aspirations are affordable healthcare, college education for kids, owning a home, having a car, saving for retirement and occasional family vacations. Planning and saving for the future are critical elements in attaining a middle-class lifestyle for most families, according to a report by the United States Department of Commerce in 2010.[15]

Some researchers define middle class based on indexes that consist of income, education, occupation, and other measurable criteria to rank people by social standing and living standards. Other social scientists define the middle class more broadly from a qualitative perspective. They are based on values and expectations that are difficult to quantify:

- Strong orientation toward planning for the future.
- Control over one's destiny.
- Movement up the socioeconomic ladder through hard work and education.
- A well-rounded education for one's children.
- Protection against hardship, including crime, poverty, and health problems.
- Access to home ownership and financial assets such as a savings account.
- Respect for the law.

The bottom line is that the income level alone is not the only criteria to describe the middle class. However, it is the most important factor in defining it in public discourse.

Shrinking Middle Class

According to a Pew Research Survey, 85 percent of self-described middle-class adults say it is more difficult now than it was a decade ago for middle-class Americans to maintain their standard of living.[16] One of the reasons is that the cost of key components affecting the middle class is growing at a faster rate than their income, which has remained relatively flat over the last twenty-five years (Fig 1.2)

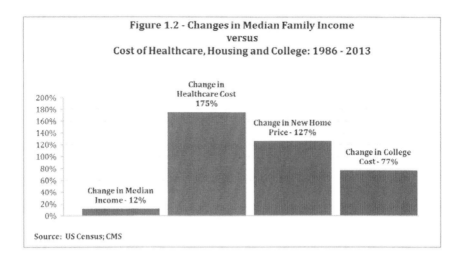

The top one-percenters have gained more over the last decade than the rest of the 99 percent. However, blaming them does not solve the problem. To think that the top one- percenters would not look out for their interests would be going against the main tenets of capitalism. One-percenters and the large corporations have the resources to finance the political campaigns not the lower or the middle class. Therefore, it should not be a surprise that national policies favor those at the top of the economic ladder rather than the bottom or the middle. Gains at the top appear to be at the expense of the middle or lower class, but that is because the economic pie is not growing. The economy, despite all of the tax breaks, government stimulation, and lower interest rates, is not growing at high enough rates to help the middle class. Therefore, the size of the economic pie remained the same and the people with influence and power have been able to get a larger share of that pie. This makes it imperative to grow the economy at a faster and sustainable rate so that individuals on all social levels can benefit.

Changing realities and shifting priorities in American society may redefine what it means to be the middle class. Some of the historical norms of owning a car, retiring at 65, or even owning a home may possibly be things of the past as millennials are not necessarily interested in buying cars or homes.[17] At the same time, the baby boomer generation may not want to retire because they cannot afford to or they want to contribute to society by sharing their experiences and wisdom.[18]

Income and Wealth Inequality

The highest one percent of income earners' average income grew by 277 percent between 1979 and 2007 and went down to 175 percent after adjusting for inflation by 2011 due to the financial crisis of 2008. During the same time, the bottom 80 percent had only 15.6 percent growth in their income, according to the Congressional Budget Office's (CBO) report of 2014.[19] Figure 1.3 shows that income gains at the top were very high compared to almost none for the middle-income households from 1979 to 2007.

One of the reasons for the income and wealth inequality is underscored in Figure 1.4. The average tax rate for the highest income earners has been going down since the 1950s from a high of 60 percent down to about 25 percent by 2010. According to a CBO report, the top two quintiles' households in 2011 earned an average income before tax of $83,300 and $234,700, and paid an average federal tax rate of 15.1 percent and 23.4 percent respectively.[20] Higher income leads to higher wealth accumulation. The top 10 percent in 2012. Most of the gains made by the bottom 90 percent during the 1960s until the 1980s have been lost over the last twenty years from as high as 36 percent in the early 1980s down to 23 percent by 2012 (Fig.1.5).

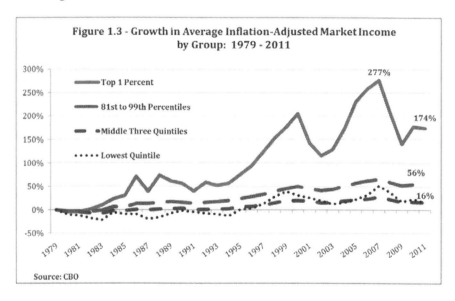

Figure 1.3 - Growth in Average Inflation-Adjusted Market Income by Group: 1979 - 2011

Source: CBO

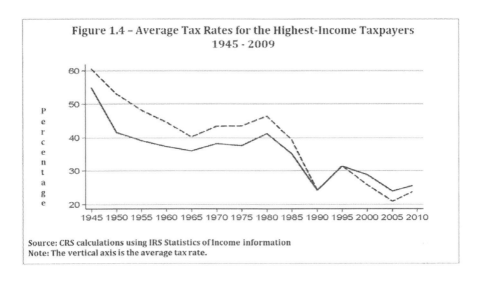

Figure 1.4 – Average Tax Rates for the Highest-Income Taxpayers 1945 - 2009

Source: CRS calculations using IRS Statistics of Income information
Note: The vertical axis is the average tax rate.

Figure 1.5 - Wealth Composition: Top 0.1% vs. Bottom 90% 1917 - 2012

Source: CBO

In addition to the tax policies, the corporate executive's compensation plan is another factor that can be attributed to a wider income inequality. During the financial crisis of 2007 and 2008, a number of companies went bankrupt or were sold in a fire sale while the executives responsible walked away with millions of dollars in severance pay and were able to keep most of their pay collected prior to the companies' failure. Some of the executives were able to keep hundreds of millions of dollars while the taxpayers were left bailing them out.[21]

In a step toward economic parity, in July 2015, the U.S. Securities and Exchange Commission (SEC) proposed rules that will require companies to adopt a clawback provision on executive compensation. The clawback provision states that executives are required to pay back incentive compensation if the results were not achieved or the company restates the financial statement or ends up in bankruptcy.

In governing publicly held corporations, the challenge is to protect the interest of the employees and the public. The best solution can be attributed to Jack Ma, the billionaire founder of Alibaba from communist China, whose governing philosophy is, "Customer number one, employees number two and the shareholders number three."[22] Representing employees and customers along with the shareholders in the corporate boards would be a step toward income parity. Jack Ma's argument is that there would be no business without customers and employees keep customers satisfied so that shareholders make money. As customers and employees gain access to the board room and become part of the high-level decision-making process, the pay gap between the employees and the executive will shrink. This is idealistic, but if a corporation founded and headquartered in a communist country and listed on the New York Stock Exchange can practice that philosophy, then American corporations can adopt it too.

Transforming Inner Cities

In addition to changing government policies on taxation and income, inner city economic growth is critical in order to bring the poor to the ranks of the middle class. The top one-percenters, some of whom are the products of the inner city, may serve as catalysts through social entrepreneurship. Reframing the inner-city narrative from crime and dilapidated buildings to an engine of growth will bring untapped human capital to society. The local, state and federal governments have been doing their part to redevelop and revive crime-ridden areas of major cities. However, what's really needed is inspiration and hope for the next generation. Many athletes, musicians, and actors were born in poverty and are giving back to their communities. Magic Johnson, Eva Longoria, Oprah Winfrey, George Lopez, and Denzel Washington are among them. They are the embodiment of the American dream and serve as roles models for future generations.

To inspire this hope, more celebrities may adopt an inner-city section, as a social experiment to empower the lower-income Americans living in those neighborhoods. The adopted section could be a zip code that represents about 10,000 people. The concept would be analogues to the adopt-a-highway idea where an organization is allowed to keep a section of the highway litter-free in exchange for placing their name on a sign on that section of the highway. It could be financially rewarding for the celebrities and, more importantly, they will be able to change lives and set the tone for transforming societies. The celebrity could earn a title of an honorary mayor for a year. However, adopting an inner-city zip code would be more than a title. The celebrity would commit one year of their life full-time to this endeavor, physically moving into the neighborhood to learn the needs of the community and showing their commitment to their section's well-being. They would bring their talents and money to bear and focus on education, business, and amplify the human talents of the community. They would invest their money, build schools, grocery stores, and whatever it takes to make their section successful as former Laker Magic Johnson has done with $500 million of properties in the inner-city areas. He employs about 3,000 people living in those neighborhoods across the country.[23]

This project would show people in the inner cities that someone cares and that education and hard work does pay off. There is no shortcut to success in life, and having a real-life successful person from their own ethnicity mentoring them would trump any government program out there.

Consumers Have the Power

The power of consumers is one reason there's hope for the middle class. It is an immense power that will show its force as excesses in healthcare, education, and other expenses grow increasingly unfair, unjustified, or take a disproportionate amount of a middle-class budget. Americans are all for competition and making more money as long as it is fair, but they do rise up when they are taken advantage of by those in power.

Consumer power manifested itself as a response to Bank of America's debit card fee in 2011. Bank of America instituted a $5 monthly fee to consumers who used a debit card for purchases. A grassroots effort by a young twenty-two-year-old woman from Washington called for a "Bank Transfer Day" where customers of big banks move their bank accounts to community banks and credit unions. She collected more than 300,000 signatures opposing the fee using the Change.org platform.[24] Upon consumer uproar and increased competition, Bank of America retreated and cancelled the debit card fee.[25]

The following year, Netflix, one of the largest online video streaming companies, decided to increase its monthly subscription rate by 60 percent. Consumers started cancelling subscriptions. The company ended up cancelling the price increase, but not before Netflix lost 800,000 customers resulting in a 77 percent decline in its stock price over four months.[26]

Consumer and social media power forced both of these large corporations to retreat. Lower and middle-class Americans were most affected by the charges, and they responded and got results. In a similar vein, there is no reason that American lower and middle classes cannot challenge the status quo and call for a transformation of policies and institutions that are undermining them. They have successfully used social media to band together to effectuate change without the government's help.

The New American Middle Class

The new American middle class is defined by the changing dynamics and shifting priorities of society. The way Americans see themselves, their possessions, their work, and their health will redefine the new American middle class.

Americans' perceptions of material wealth are changing. According to a 2011 survey, only 50 percent of consumers agree with the statement that, "owning things is a good way to show my status in the society." Happiness studies show that experiences increase contentment far more than purchases do, and young people's intrinsic understanding of this is fueling an experience economy.[27] The idea of buying a car, moving to a suburb, and owning a home may not be the middle-class aim of the future. The emerging model of sharing and monetizing the underutilized assets such as cars and homes may reduce the drive to purchase those assets, which can turn the purchasing behavior from the previous generation upside down. According to a report by PWC, one of the largest multinational professional services company, 86 percent of the adults familiar with a sharing economy agree that it makes life more affordable, and 81 percent agree that it is less expensive to share goods than to own them individually.[28]

The young generation appears to be more focused on quality of life and value staying close to family and friends, having free time for recreation and working in creative jobs.[29] The millennial generation that represents over 25 percent of the population is less likely to be homeowners or even car owners than young adults in previous generations.[30]

Technological advances in daily life and workplaces create efficiency that benefits the economy. While this eliminates some jobs, the emerging and re-engineering of legacy industries create better and higher paying ones. This paradigm shift toward a 1099 economy, on-demand employment and services, declining marriage rate, delayed childbearing, disincentive to own a car, and other evolving trends will define the future. Indeed, even the historical definition of family purchasing power doesn't reflect the changing social, cultural and economic shift taking place in society. For example, value created by Google or Facebook are not accounted toward the national income since they are not purchased services.[31]

Middle-class households spend a higher proportion of their income on consumer goods than upper-class households. Hence, higher consumer spending due to lower oil and commodity prices, lower interest rates and an improving job market will begin a new American middle class.

Finally, citizens have learned from the difficulties of the previous decade. A significant portion of young adults are working through student debts, and those who lost their houses during the 2007 housing crisis are beginning to get back on their feet. Their demands for accountability and transparency in education, healthcare, housing, and tax codes will lead to the middle-class comeback—starting with women leading the way.

Two

Women Leading the Way

Over the last two decades women in the United States have made great strides toward professional and financial advancement. Women's role in every aspect of the American life is growing and will play a critical role in shaping the future. Over the last thirty years, women have made up an increasing share of the workforce.[32] Wage inequality is decreasing. Women are projected to surpass men in wages as more of them earn college degrees[33] and their value and contribution to the workplace is fully acknowledged and appreciated.

Some of the most promising signs are in the following areas:

- Women account for almost half of the labor force, and they earn 79-cents for every dollar their male counterpart (for the same job)[34] makes and are gaining ground.
- 50 percent of the law and medical degree earners are women.[35]
- Contribution of wives' earnings to family income has increased from 27 percent in 1980 to 37 percent by 2011.[36]
- Seventy-one percent of women who graduated high school in 2012 enrolled in college following graduation compared to 63 percent

in 1994. At the same time, it remained unchanged for men at 61 percent.[37]

- The number of Fortune 500 women CEOs reached a historic high to 24 in 2014 compared to only one in 1998.[38]
- Full-time working women contribute more than half of their household income on average compared to 25 percent a generation ago.[39]
- Women's median hourly wages increased by 50 percent or more over one generation.[40]
- A record 104 women (83 in the House and 21 in the Senate), in the 115th Congress as of January 2017, compared to only 11 (10 in the House and 1 in the Senate) in 1970.[41]

There are more self-made women billionaires than before, such as Diane Hendricks of ABC Supply, Jin Sook Chang of Fashion 21, Judy Love of Love's Travel Stops and Country Stores, Marian Ilitch of Little Caesars Pizza, and Sara Blakely of Spanx. Some of the more visible women corporate leaders are Sheila Johnson of BET Networks, Mary Barra of General Motors, Indra Nooyi of Pepsi, and Sheri McCoy of Avon Products. These women, and many others, are not only leading the way toward gender equality in the executive boardroom, they are also mentoring the potential future women executives.

On the political front, former Senator Hillary Clinton became the first woman to clinch the Democratic Party's presidential nomination for the 2016 election. This was the first time in United States history that there was a woman contender for a presidential election. She won the popular vote by a wide margin but lost the election to Donald Trump. After her campaign for the Democratic nomination in 2008, Clinton remarked, "Although we weren't able to shatter that highest, hardest glass ceiling this time, thanks to you, it's got about 18 million cracks in it (referring to 18 million votes cast for her) and the light is shining through like never before, filling us all with the hope and the sure knowledge that the path will be a little easier next time."[42]

On the congressional and judicial front, the second and third branches of the federal government, House Minority leader Nancy Pelosi was the first woman to serve as Speaker of the House from 2007 to 2011, and she is considered the highest-ranking woman in congressional history.[43] The current Supreme Court has Ruth Joan Bader Ginsburg, Elena Kagan, and Sonya Sotomayor, the highest number of women ever to serve on the court; in fact only one other woman, Sandra Day O'Connor, has ever served as a Supreme Court Justice.[44] All of these leaders are paving the way to for the next generation of woman executives, politicians and justices, among others.

Almost 50 percent of the law and medical degree earners were women in 2010 compared to only 10 percent in 1971.[45] Today more women than men are going to college and graduating.[46] According to a Pew Research Report, women's median hourly wages increased by 50 percent or more over a generation.[47] Yet, women still make less than men for the same job. However, American society is making strides in a positive direction: women made 60 cents for every dollar men earned in 1980, but the figure rose to 79 cents by 2014 (Fig 2.1).[48]

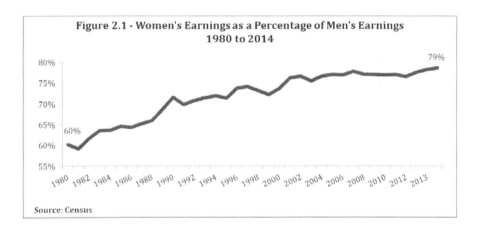

Technology and a shift in social attitude have made it easier for women (and men) to balance family and a career. Women control 80 percent of consumer spending in the United States,[49] and as a result, the importance of women in a family's financial affairs will grow as gender equity increases in all aspects of society.

Education

Education is vital for economic prosperity for all, women or men. Education will not only improve women's chances of gaining parity with men in the workforce, but also allow them to lead public or private institutions.

Women have surpassed men by two to one in earning four-year college degrees from 1980 to 2013, according to a report by the Institute of Education Sciences, part of Department of Education (Fig. 2.2).[50] Moreover, women outpace men in college enrollment, especially among racial and ethnic minority groups. According to a report by the Pew Research Center; in 2012, 71 percent of women enrolled in college right after high school compared to 63 percent in 1994.[51]

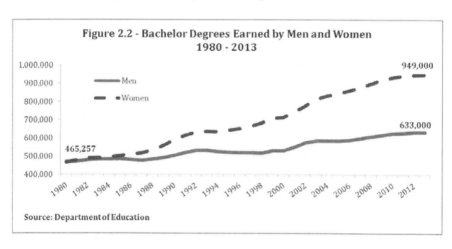

Figure 2.2 - Bachelor Degrees Earned by Men and Women
1980 - 2013

Source: Department of Education

Women with college degrees can command a higher compensation, especially as they make up a larger share of current graduates. It will considerably enhance their chances of getting higher paying jobs as the supply of male graduates is declining in relation to female graduates. This more equitable career trajectory leads to more and more women in management positions, which in turn will improve women's power within an organization and increase their purchasing power. According to a study by Catalyst, a nonprofit that addresses topics vital to women in the workplace, as of 2013, women in management, professionals, and related occupations represented 51.4 percent of the United States workforce. More women in the workforce will lead to more women in executive positions.[52]

Political Engagement

Diversity is one of America's greatest assets, unparalleled anywhere in the world. It is ironic that the world's sole superpower and champion of democracy is not fully utilizing its assets by not having women as a full partner in public deliberation.

Women are well underrepresented when it comes to legislative bodies. The United States is ranked ninety-first in the percentage of women in the national legislatures, according to a report by American University.[53] Only one in five Senators and Representatives are women, majority of them are Democrats while Republican women make up only six percent of the Senate and five percent of the House (Fig. 2.3).[54]

In fact, twenty-three states never had a woman governor, including California, Illinois, and New York to name a few. In other words, at least one hundred million Americans have never been represented by a woman governor. There has never been an African American woman governor and there has only been one Latina governor in United States' history. This ratio can only get better as women's participation in politics continue to grow. One of the reasons for that optimism is that 24 percent of the state legislatures are women compared to only 5 percent back in 1971. State experience can serve as a catalyst for a national position.

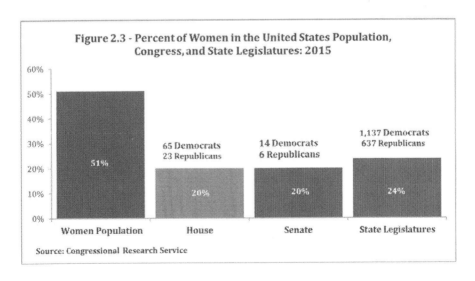

Figure 2.3 - Percent of Women in the United States Population, Congress, and State Legislatures: 2015

One adverse outcome of this underrepresentation of women is that the United States' foreign policy relies on military force to solve international disputes. As former Secretary of Defense to Presidents Bush and Obama, Robert Gates[55] said that America's most recent presidents, "have been too eager to pull a gun to solve a problem." According to a survey of the forty-three women politicians around the world by *Foreign Policy* magazine, 65 percent[56] agreed that "the world would be more peaceful if more women held political office." The author continues, When I ask my peers about this gap, three consensus opinions emerge: 1) Women are more likely to see the other side's point of view, 2) Women are less likely to see the world as a zero-sum game, 3) Women are more likely to believe that bombing someone does not ultimately achieve anything.

The United States' fifteen-year long military engagements in Iraq and Afghanistan support this thought that bombing does not win the hearts and minds of people.

Corporate Engagement

Research shows that companies with higher representation of women on the board or in top management exhibit higher return on equity, higher valuation, and higher payout ratio. Overall, a higher representation of women leads to better financial performance.[57]

There is no question that a glass ceiling still exists and that it prevents women from reaching top positions in business and politics, but it is showing some cracks, and those cracks will continue to grow as more and more women become managers, business owners, and entrepreneurs.

Women make up a very small percentage of the Fortune 500 CEOs. A broad spectrum of companies have CEOs who are women, including General Motors, Hewlett- Packard, IBM, Pepsi, Xerox, and Avon as of 2014. These CEOs and other senior level executives serve as role models as well as mentors to the next generation of women aspiring to leadership. This may accelerate the pace of women's participation in executive rooms. Yes, it will take a while to gain equality in proportion to women's share of the population overall, but it is a positive sign.

The manufacturing sector employed about twelve million workers in 2013. However, the industry is struggling to find skilled workers. The projected economic expansion and baby boomers retiring will further exacerbate the shortage. Since women are graduating from college at higher rates than men, they will have an opportunity to increase their representation in this high-paying sector.[58] Women are the largest pool of untapped talent for the manufacturing sector since they make up only about 25 percent of the manufacturing workforce.

Closing the Income Gap

A generation ago, women were relegated to menial jobs and subjected to discrimination and sexual harassment. In 2012, women made up almost half (47 percent) of the United States labor force, up from 38 percent in 1970 with more management positions. This is getting closer to their population representation.

Women will also benefit as many industries face a talent shortage, as aging workers retire from industries dominated by men, and the economy becomes more dependent on skilled workers. As a result, women will gain parity with men's wages, and they may also surpass men as both genders gain an equal voice in the boardroom.

Despite the income gap, women working full-time contribute more than half of their household income today compared to 25 percent a generation ago.[59] Historically, one of the factors for wage gap has been that many women were likely to work in so-called "pink-collar" jobs, such as housekeeping, service, and caregiving occupations where they were underpaid, even though this sector was among the fastest growing in the United States. Another contributing factor may be that women were more likely to take time off work when they had children or family demands, or they worked part time.[60]

A record 40 percent of all households with children under the age of eighteen include mothers who are either the sole or primary source of income for the family, according to a new Pew Research Center analysis of data from the United States Census Bureau.[61] The share of mothers saying their ideal situation would be to work full-time increased from 20 percent in 2007 to 32 percent in 2012 while those saying they prefer not to work at all fell from

29 percent to 20 percent. The survey also found that the total family income is higher when the mother is the primary breadwinner.[62] The wives' share of family income had grown from 27.5 percent in 1971 to 37.5 percent by 2011 (Fig. 2.4).[63] All of these factors will empower women with more purchasing power and pave their way toward their middle-class comeback.

Figure 2.4 - Contribution of Wives' Earnings to Family Income: 1971 - 2012

Source: BLS Reports - December 2014 - Table 25

Women Entrepreneurs & Empowerment

Women control $11.2 trillion, or 39 percent, of the nation's investable assets as decision makers, not just influencers.[64] They make the final investment decisions. More women are getting business degrees—16 percent in 2011 compared to 9 percent in 1970.[65] As of 2009, firms owned by women created or maintained more than twenty-three million jobs with an annual economic impact of the $3 trillion in the United States [66]—that's equivalent to the fifth largest economy in the world if the women-owned businesses were their own country, ahead of France, the United Kingdom, and Italy.

Many talented women are forging their success outside the established corporate world by starting their own companies. According to a study by Catalyst, a nonprofit research and advisory organization, 51 percent of women left the corporate world due to their desire for more flexibility.[67] Between 1997 and 2014 the number of women owned businesses grew by 68 percent, one-and-a-half times the national average.[68]

Julia Hartz, who worked as a TV executive, cofounded Eventbrite, a company with 500 employees. Their platform allows people to organize events and purchase tickets. Its platform is available in 190 countries. She was recognized as one of *Fortune* magazine's "Most Powerful Women Entrepreneurs" in 2013. Caterina Fake founded the photo-sharing website Flickr after working for Salon.com. It was sold to Yahoo in 2005 for a reported $35 million.[69]

Others are starting nonprofits to empower individuals. Jessica Jackley cofounded Kiva, a nonprofit with a mission to connect people through lending to alleviate poverty. People can lend as little as $25 to help create an opportunity for loan-seekers in the United States or around the world through a worldwide network of microfinance institutions. Since its founding, Kiva has over a million lenders with almost $800 million in loans in eighty-three countries. More than 80 percent of the loans were made to women with a 98 percent repayment rate.[70]

Technology is one of the key factors that is and will continue to empower and enable more women to achieve middle-class status and elevate it as a whole. Women are no longer limited to a physical office or nine-to-five jobs. They can work from home at any time of the day or night and set their hours. With a laptop, smartphone, and Internet connection, anyone can work as a call center professional, computer programmer, or manage a business—and women are taking advantage of these opportunities.

Three

Technological development will be one of the key catalysts that will help with the middle- class comeback by introducing competition and efficiency to some of the age-old operational models in education, healthcare, housing, and taxing. It will squeeze excess capacity in transportation, lodging, logistics, buildings, and other sectors alike and turn them into productive, revenue generating ventures. It will eventually lead to less waste, increased capacity, and shorter delivery time of products and services. This chapter is focused on four key areas that will have a larger positive influence on the middle-class pocketbook due to technology: transportation, energy, environment, and communication.

Technology is another area where women are gaining parity, even though there is still a long way to go. However, Sheryl Sandberg of Facebook, Meg Whitman of HP Enterprise and former CEO of eBay, Neerja Sethi of Syntel, Marissa Mayer of Yahoo, Susan Wojcicki of YouTube, Ursula Burns of Xerox, Virginia Rometty of IBM, and many others all represent hope and change in the technology field and serve as role models for other aspiring women leaders.

Technology is reshaping American culture and the economy like never before. America is moving away from a centralized economy to a distributed and sharing economy. The United States is generating electricity through solar panels as part of a distributed energy platform; information sharing and data processing have been decentralized through tablets and smartphones; Uber and Lyft are allowing car sharing through technology; Airbnb and HomeAway. com allow people to share physical space; businesses and individuals are sharing, storing, and processing data through cloud-based services. All of this re-envisioning will continue to reduce costs and make us more productive, which will result in increased purchasing power for average Americans.

It appears as if the United States is going back to the good old days of decentralization. Daily necessities are delivered to us, doctors are making house calls, farmer's markets are cropping up in every locality, and electricity can be generated in homes. Americans are living in an information age, and the digital revolution is underway. Not only will technology reduce the cost of education, healthcare, and housing, but it will change transportation, use of energy, environment, food distribution, and communication. All of these factors will reduce the cost of living for all segments of society, but most importantly for the lower and middle class.

Over the last two decades some of the gains in standards of living can be attributed in part to increased productivity through distributed platforms. A Distributed Ecosystem can be used as a metaphor for the distribution, collaboration, and sharing of knowledge, production and services. The Internet is the critical medium for this evolving ecosystem and will continue to transform American society and culture. These platforms will continue to evolve and make industries more efficient and improve Americans' quality of life as long as the benefits are shared by everybody, not the top one-percenters.

Transportation & Mobility

Almost every aspect of life and the economy depends on transportation, and improving transportation through technology can greatly benefit the American middle class. The auto sector alone accounted for revenue of $2 trillion in 2014 that includes auto manufacturers, rentals, oil and other related factors.[71] The current cost of approximately $1 per passenger mile may come down to around 30 cents due to higher rates of asset utilization as the new technologies are implemented.[72] The new transportation model will even change Americans' behavior in terms of car ownership. Companies like Uber and Lyft, autonomous cars, and drones in the air will shape the future of transportation over the next decade.

The current passenger car model for a working person is to take a car to work alone; the car sits idle for rest of the day in a parking lot for eight hours or longer, and then the person drives it back and again it sits in the garage for the night. Instead of sitting idle for hours, technology may eliminate the idle time by making better of use those cars. Experts see the autonomous-car revolution as a beginning,[73] which would lead the way to a self-driving or driverless car.

In the meantime, Uber and other similar companies minimize the downtime a car is sitting idle. Using Uber can be more cost effective for riders by eliminating parking and gas cost, and saving on car insurance and depreciation. It's all done via the click of an icon on a smartphone and within three to five minutes a car is at the door—much faster and affordable than a taxi. As a result, millennials are less inclined to own a car.[74] In cities, it saves parking spaces that can be used for more productive use such as a warehouse, housing, or community centers as well as reduce environmental harm.[75] As competition grows, these services will become even more cost effective for riders.

Alternatively, an Uber driver has the option to service a particular ride based on his or her financial needs and location. This model allows every person who owns a car to be an entrepreneur and a freelancer, selecting his or her own work schedule and location.[76] Millennials and urbanites are moving away from a consumption model of ownership to a pay-per-use concept, eliminating the expense of owning a car—about $0.97 per mile, which includes depreciation, financing, insurance and fuel cost.[77] Moreover, more efficient use of cars will reduce harm to the environment. Technological breakthroughs in the driverless car will not only be a reality for personal use but also for commercial use.

Technology is also giving businesses better and less expensive models for delivery of their goods. Amazon's goal is to provide same-day delivery to its customers via its own network of delivery couriers through a program called Amazon Flex. In addition, they aim to deploy drones for delivery of small packages in the next five years. Eighty-six percent of their packages in 2012 weighed five pounds or less.[78] Tech-driven delivery models would decrease traffic congestion and the need for parking spaces, while reducing the cost of goods for businesses and consumers.

The auto industry is being transformed from being just the manufacturer to the provider of mobility by integrating technology (GPS, Internet, and phone) into their cars. This will help consumers make better use of their travel time to work, communicate, and rest.

With the aging of the American population, driverless cars could provide freedom and independence for those with vision or other health impairments that would prevent them from safe driving. Not having them drive will not only save lives but will also reduce overall societal cost. They would not have to depend on their children or other support persons to take them back and forth for their daily activities or visits for medical care.

Safety, of course, is not just a concern for the elderly driver. About 13,000 deaths can be attributed to alcohol at a cost of $37 billion every year.[79] Reducing accidents and deaths due to drunk driving will be one of the greatest benefits of driverless cars. Even today in 2017, smart cars can detect the alcohol level of a driver and can refuse to start. This gain in safety also reduces insurance and other costs, which benefits middle-class Americans.

Auto manufacturers like Daimler of Germany are working on a driverless truck that would reduce accidents and save lives in commercial driving. Human errors were the major cause of 330,000 large trucks that were involved in crashes that killed 4,000 people in 2012. Driverless trucks could prevent more than 70 percent of those crashes.[80] The estimated cost of truck and bus crashes to the United States economy was $99 billion in that year.[81]

Daimler was granted the first license by the state of Nevada to test such vehicle on a public highway. In fact, fifty self-driving trucks are being used in Australian mines. In addition, Lockheed Martin has built a range of autonomous trucks for the United States Army. According to the American Truck Association, United States trucking companies could see a shortfall of 240,000 qualified truck drivers by 2022.[82] That shortage could be compensated through driverless trucks. Most of these savings and efficiencies could be transferred to the consumers in lower costs.

Energy

Technology can also benefit the energy sector, and through it, the middle class. The solution to cheaper and environmentally friendly energy is on roofs, roads, farms, and cars.

American households with a pretax annual income below $50,000 (a group that makes up almost 50 percent of the United States households) spend about $3,893 per year on residential and transportation energy.[83] That cost highlights the potential benefits of the many renewable and alternative energy sources available. Rethinking energy sources could make America self-sufficient as a country and reduce a middle-class family's expenses.

The United States is projected to be energy independent by 2020 due to the oil and gas production started with hydraulic fracturing. By some accounts 30 percent of oil price reduction during the summer of 2015 can be attributed to the growing United States oil production.[84] In the meantime, energy productivity has increased by 54 percent since 1990 and gasoline consumption has been declining due to consumer preferences, a declining number of cars on the road, and fewer driven miles per vehicle.[85] Furthermore, renewable energy options are getting more economical and are environmentally friendly.

The information and digital revolution can help America move away from the centralized electricity delivery model to a decentralized one. Oil and coal have been the major energy resources over the last few decades. However, the evolving models in transportation and mobility will significantly reduce the use of oil by making vehicles more efficient. One of the models is Distributed Energy.[86]

Distributed energy models feature smaller devices to provide electricity locally to consumers. They can be part of renewable energy such as solar, wind, micro-turbines, and fuel cells. Not only are they cleaner, but they will also make building multibillion-dollar power plants a thing of the past. Every residential, commercial, and industrial roof and even the roads can produce energy. As technology develops, the outer walls of a building could even be generating electricity through renewable energy components built into the paint.[87]

On the auto front, a new Mercedes-Benz concept car will not only be powered by hydrogen but also have a multi-voltaic silver paint that harvests solar and wind energy to power the car.[88] Even more fascinating is the first solar roadway laid out in the Netherlands that generates solar power through embedded solar panels. As technology develops every surface exposed to the sun is a potential source of energy.

Vehicle-to-grid (V2G) is a system being tested where the power stored in a car can be transferred to the centralized power grid, providing a revenue source to the owner. Since 95 percent of the cars are parked at any given time, batteries in electric vehicles can let electricity flow back from the car to the electric grid and back depending on the needs of the grid operator. This could be worth an estimated $4,000 per year in cost savings to the utilities, a part of which can then be paid back to the driver of that electric or hybrid car.[89]

Moreover, the industrial and transportation sectors accounted for almost a 50 percent share of total energy consumption in 2014.[90] They are projected to reduce their energy consumption and operational costs through driverless and hybrid cars and trucks. All these innovations impact America on a large scale and individual households; on both levels, the savings benefit the middle class.

Environment

A clean environment will improve the health and increase the lifespan of average Americans, and this helps the middle class by reducing medical costs and enhancing the quality of life. American ingenuity will lead the way in reducing the negative impact of greenhouse gas through technological advances such as the distributed energy model, deep-sea bacteria that could neutralize greenhouse gas, zero-net houses, and others.

Fossil fuel is used for 67 percent of the electricity generated in the U.S. and the renewable energy accounts for only 7 percent.[91] The environment will benefit from the shift from fossil to renewable energy. According to one recent finding, Miami, New Orleans, and other cities may sink below sea level by the end of this century if the trend in global warming continues.[92] Renewable energy production will grow as technological advances such as distributed energy, and zero-net houses become prevalent. A zero-net house is defined as one that generates enough renewable energy to sustain itself. This will reduce the number of power plants using coal by 50 percent as they age and become obsolete or less profitable. Coal plants are the nation's top source of carbon dioxide (CO_2) emissions, the primary cause of global warming.

Evolution of 3D printing will bring manufacturing to point of purchase, which will reduce the transportation cost of finished products resulting in reduced energy use. Emails and text messaging have significantly reduced the need for postal mail thus saving trees due to the diminished use of paper and gas due to the reduced number of physical mail deliveries. Electric, hybrid, and hydrogen cars will reduce the use of gas and the need for new refineries. Also, the miniaturization of products, such as a smartphone that has more power than PCs of ten years ago, LCD TVs replacing old bulky models, or computer monitors, has reduced the shipping costs. It will continue to reduce gas and the packaging material costs required to deliver them.

Willow and Gorilla Glass, a revolutionary product under development at Corning, will have the flexibility of plastic and durability of steel that would ship on a roll just likes plastic.[93] It could potentially replace tablets, TVs and other viewing devices that could be folded and placed in a pocket to be opened later just like a handkerchief. It could replace house or car windows, which again will reduce the cost of delivery and storage while increasing the life cycle of those products.

Communication

The advent of the Internet, laptops, smartphones, and other digital devices has forever changed social, family and business relationships. Instant communication is tailor-made for Americans' need for instant gratification. One of the major accomplishments over the last two decades has been the evolution of communication technology that has challenged the existing culture and social norms of daily life and interactions. The word multi-tasking sums up how Americans function in this new age, emailing, texting, chatting, listening to music, Tweeting, and performing other tasks all at the same time. This has become one of the most productive generations in human history due to incredible advances in the communication field.

In the past, making an overseas call was time consuming and expensive, up to $5 per minute to some countries. Now it costs as little as five cents per minute or even nothing with no waiting. Smartphones are more important than even cash for performing daily tasks and chores—finding a phone number or directions, sending documents, or researching products or ailments. A person can even watch their house 24/7 from anywhere in the world.

Technological developments in communication improves the quality of life and productivity in so many ways that it is hard to quantify its total impact, let alone the overall benefit to family budgets in time saved. There is a paradigm shift in communication through social media and other platforms that were not imagined even two decades ago that will ultimately benefit the middle class. That is where millennials become a new rising force in the public and private sector.

Four

Millennials

Millennials are exerting a significant influence on the future when it comes to education, healthcare, and housing because they make up more than a quarter of the total United States population, or about 85 million people. They carry an estimated $200 billion of direct purchasing power and $500 billion of indirect spending.[94] That power will continue to grow as they age to peak buying power in their forties and fifties, with a projected $30 trillion in inheritance[95] over the next several decades. Combine that with about $1.5 trillion in investable assets,[96] and they will shape the economy and public policy for decades to come through new venues of communication by voicing their concerns and acting upon it.

The White House report describes millennials as the generation born between 1980 and mid-2000s[97] and the Pew Research Center as one born between 1981 and 1996[98]. There is no definitive agreement on the dates. Regardless, they will continue to be a significant share of the population for decades to come as the baby boomer's size declines. They are the most diverse and educated generation to date with 44 percent being part of a minority race or ethnic group.[99]

They have witnessed the dot-com bubble, followed by the housing and financial crisis, and two wars all within their lifetime. Despite all of that turmoil, they are more optimistic than older adults and believe that the country's best years are ahead of them.[100] They are redefining culture through social media. About half to two-thirds of millennials are interested in entrepreneurship, and more than 27 percent are already self-employed.[101]

Millennial women are the first ones in modern history to start their working lives with near parity with men in terms of earnings, according a survey by the Pew Research. In 2012, among workers ages twenty-five to thirty-four, women's hourly earnings were 93 percent of men's.[102]

Some of the well-known, successful millennials are Mark Zuckerberg of Facebook, Joe Gebbia of Airbnb, Taylor Swift the singer, and Evan Spiegel of Snapchat among others who are reshaping and creating new categories of businesses and industries.

Reshaping & Creating Industries

Millennials are transforming existing industries while creating new ones that were not even imagined a few years ago. Whether it is music, communication, media, or consumer goods, they are creating jobs, enhancing consumer's purchasing power, and improving the quality of life for all Americans.

Facebook gathered more than one billion users around the world in ten years—if those users were a country it would be the third largest in the world. It has changed how people communicate, share information, define friendship, engage in politics, view and use media. In addition, Facebook has created 4.5 million jobs in 2014 with $227 billion in economic impact globally and $100 billion and one million jobs in the United States.[103] It can claim some credit for the election of then-candidate Barack Obama to be the President of the United States in 2008.[104] Twitter on the other hand was the main communication medium for Donald Trump in his rise to the presidency of the United States.

Facebook has made it easy to get news from around the world and to converse and collaborate with people in many different countries. This platform along with Twitter was one of the information sources and organizing vehicles in Tunisia, Egypt, and Libya[105] during the rise of the Arab Spring. Through its Instant Article, a publishing platform and in partnership with The New York Times, Buzzfeed, NBC, and others, it serves as a vehicle for readers to share articles with friends and family.[106] Thus it is becoming one of the largest platforms for news and information.

Danae Ringelmann, a borderline millennial and a daughter of a small business owner, is taking on the traditional top-down model of bank financing and giving people, rather than bankers, the chance to decide what projects get funded. She cofounded Indiegogo,[107] which hosted 250,000 campaigns in 224 countries and became synonymous with crowdfunding. The concept of selecting worthy causes to fund is popular among the millennial generation. Indiegogo's platform has helped an African humanitarian get a kidney transplant and a couple who needed an expensive IVF treatment to have a child.[108] Misfit Shine, an elegant wireless activity tracker designed by Misfit Wearable raised nearly $850,000 on Indiegogo in 2013. The company then went on to raise over $50 million more through venture capitalists and now sells products at Best Buy, Target, and Apple stores.[109]

Taylor Swift, who was worth $200 million at age twenty-five, became the agent of change by calling for fair treatment of artists by refusing to provide free access of her album to Apple. Swift, who has sold millions of albums around the world, took Apple, one of the most powerful companies in the world, to task in defense of intellectual property rights. She challenged Apple's policy of not paying royalties to artists for any streaming of their songs during subscribers' three month free trial. She withheld her latest album *1989* from Apple's streaming service and said on Tumblr, "We don't ask you for free iPhones. Please don't ask us to provide you with our music for no compensation." By that night, Apple changed its policy and began paying royalties to artists during three-month free trials.[110]

Airbnb, cofounded by Joe Gebbia, Brian Chesky, and Nathan Blecharczyk, has revolutionized the lodging industry in just eight years. It has created a community marketplace to list, discover, and book accommodations around the world. People can book an apartment for a night, a castle for a week, or a villa for a month in 190 countries and more than 34,000 cities at a price that fits their budget. In the process, they have revolutionized the lodging economy and created additional sources of revenue for millions of people who have extra space to spare.

Evan Spiegel and Bobby Murphy cofounded Snapchat in their mid-twenties. The video and photo messaging app that has more than one hundred million active users and more than five billion video views per day has the fastest growing user base. User images are available for only a very short duration of about ten seconds, and then they're deleted. In December 2015, the app entered into real-time news when it allowed people near the attack in San Bernardino, California to show what they were seeing. Instead of stories filtering through the media, it was reporting live experiences of people being witnesses to an event.

Some of the millennials, with their money and celebrity power, are engaged in empowering others through philanthropy and social entrepreneurship.

Philanthropy & Social Entrepreneurs

The millennials are transforming philanthropy, from sending checks to texting a donation via their phones or social media platforms. Millennials, like the generations before them, are interested in helping others and making a difference, but they're reinventing the methods and means to show their support, both empowering and directly benefiting the middle class.

The Ice Bucket Challenge that started in the summer of 2014 raised $220 million for ALS Association and became the world's largest global social media phenomenon.[111] This virtual chain letter was started by a millennial: people hoisted buckets filled with ice water and poured it over their heads and nominated a few friends to do the same to raise awareness of ALS, commonly known as Lou Gehrig's disease, and make donations. The challenge spread like a wildfire and ended up all the way to the White House along with many celebrities and public figures participating.

Warby Parker, a for-profit company, cofounded by David Gilboa, is bringing philanthropy to consumers. Through its "Buy a Pair, Give a Pair" program, for every pair of eye glasses purchased, the company donates a pair to a person in need. It has distributed one million pairs throughout the world to those who need but cannot afford glasses.[112]

Mark Zuckerberg sums up the attitudes of many millennials as they enter parenthood and begin thinking of the generations that will come after them,

"Like all parents, we want you to grow up in a world better than ours today … We will do our part to make this happen, not only because we love you, but also because we have a moral responsibility to all children in the next generation." In Zuckerberg's letter, he and his wife Pricilla Chan pledged to donate 99 percent of their Facebook shares, currently worth $45 billion, during their life to advance that goal.[113] In 2010, he had donated $100 million to Newark, New Jersey's failing public school system.

Political & Social Engagement

The millennials will represent the largest voting bloc in the next few elections. About half describe themselves as politically independent.[114] They are social liberals and fiscal centrists, according to a survey by the Reason Foundation.[115] Their voices will shape the future of the country, particularly for the middle class.

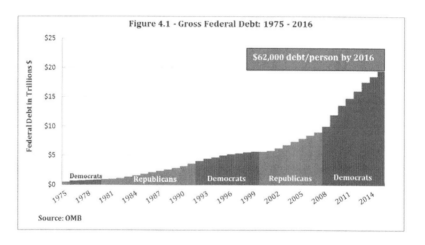

Millennials face trillions of dollars of federal debt (Fig. 4.1) that previous generations continue to accumulate, and there is no plan to balance the budget any time soon. As millennials see increasing taxes as their incomes rise, this idealistic, politically vocal generation will call for a change and will have the numbers to effectuate a transformation of the political system.

"Millennials will produce radical reconstruction of civil institutions and government," according to Michelle Diggles, an advisor at a Democratic think tank Third Way. As one of the high school students responded to a reporter's question about public service, "Let me tell you what's going to happen to government and politics when we get hold of them. We will destroy them."[116] They will do this through political entrepreneurship and disruption.

Black Lives Matter (BLM) was cofounded by millennials Alicia Garza, Opal Tometi, and Patrisse Cullors. It started in the aftermath of the acquittal of George Zimmerman in the 2013 shooting death of Trayvon Martin, an African-American teen in Florida. The movement started with a simple hashtag, #BlackLivesMatter on Twitter. It mushroomed into today's civil rights movement through the power of social media,[117] whether one agrees with them or not. According to their website, it is "an ideological and political intervention in a world where Black lives are systematically and intentionally targeted for demise. … It is an affirmation of Black folks' contributions to this society, our humanity, and our resilience in the face of deadly oppression."

The BLM has 26 chapters that largely set their own direction. They are engaged in various ways from protesting to organizing, to holding rallies, and calling for more minority vendors at the state fair. The decentralized structure of the movement helps local chapters customize solutions to local problems. It served as a catalyst for activism on college campuses, such as the University of Missouri in 2015. R. Bowen Loftin, the chancellor, was forced to resign in reaction to students' protests toward the administration's tepid response to racism on campus.[118]

Ben Rattray cofounded Change.org with a mission to empower people everywhere to create change they want to see in the world. It combines the vision of a nonprofit with the flexibility of a tech startup. It is an online petition tool used by more than 100 million people to bring about change in their communities, country, and the world. The petitions address education, environment, health, politics, and human rights.

One of the biggest concerns for the millennials is disconnect between what they learn in colleges and the skill sets required to get a high paying job. Jobs are also a big concern for lower and middle-class working Americans in order for them to obtain a higher standard of living. Even though the jobs abound, millions are not qualified to get them.

Five

Jobs and the New Economy

The United States had 5.5 million job openings in January 2016, according to the United States Bureau of Labor Statistics.[119] That means employers could not find the right people to hire. The Bureau of Labor Statistics (BLS) projects 2.3 million computer-related professionals and engineers will be needed by 2022 to meet the demand of the economy and to replace the retiring workforce.[120] There will be a shortage of 175,000 truck drivers by 2025.[121] Furthermore, Boeing projects more than 550,000 new commercial airline pilots will be needed over the next two decades, or about 28,000 per year.[122] That places the skilled and talented labor force in the driver's seat to ask for higher compensation.

Much of this employment gap can be attributed to jobs that require science, technical, and vocational skills to perform newly-created digital jobs. This skill gap is a reflection of disconnection between institutions of higher education and the employers that need to fill those openings, as discussed in Chapter Fourteen.

Companies in Silicon Valley are clamoring for prospective employees by enticing them with perks. At Facebook, employees can pick anything from vending machines (snacks, drinks, computer accessories), are provided three meals a day five days a week, have an onsite barbershop, dry cleaners, and bikes, all at no cost.[123] Not to be outdone, LinkedIn offered unlimited vacation and seventeen paid holidays to its employees starting in November 2015. The discretionary time off (DTO) model allows employees to work with their managers to plan vacation time with no limit to the number of days.[124] In 2015, Amazon started offering twenty weeks of paid time-off for new moms while Netflix offered as much as one year paid leave for new parents.[125] All of these perks are offered without government's interference or mandates. This shows the power of competition for talents and what a growing economy can do for the middle class.

When middle-class Americans feel secure in their jobs, they can plan on buying new clothes for their kids, a computer, a car, etc. It will then results in demands for additional products, at which point businesses jump in and build the production capacity to sell those products. The net outcome is additional jobs and more economic stability for workers.

The first and foremost factor in job creation is stable and consistent business policies including regulations and the tax code. In addition, the country's infrastructure, skilled labor, and local demand for the products are also critical factors in determining whether or not a business will make a particular investment.

The simple logic behind business decisions is that businesses invest when it is a good investment—there is a demand for their products, and they can make a profit producing and selling their goods or services. Businesses don't avoid investing just because taxes are high. Taxes are irrelevant if a business is not making a profit. Billionaire Warren Buffett said raising taxes on the wealthy won't stop them from investing, and he called on policymakers to boost income tax rates for people whose income is more than $500,000.[126]

However, lower taxes are a good incentive when competing in a global economy. In the Internet era of high technology and globalization, there is nothing permanent about jobs. The days of lifetime employment don't exist anymore—even China no longer has the "Iron Rice Bowl" policy, a Chinese term referring to a permanent job.[127] The only chance for continuous employment is to be a lifelong learner. Skill sets instead of experience will be the key factor in getting jobs because of ever-changing technology. Skill sets provide a more comprehensive view of one's ability to learn and adapt in all facets of life including communication while experience reflects work history. No one can predict job prospects in 10 years because a substantial percentage of future jobs have not even been invented yet.

Companies will set up manufacturing plants wherever the demand is the greatest to meet consumers' desire to acquire them quickly. American companies have been going overseas, and foreign companies are coming to America and establishing manufacturing facilities here—to be close to the people who want to buy their products. Toyota, Honda, BMW, and Mercedes produce cars in America and even export some to other countries from the United States.[128] Airbus, a European consortium that makes the A320 aircraft and competes with Boeing, is planning to make its planes in Mobile, Alabama, to bring products closer to their consumers, creating 1,000 high paying jobs by 2017.[129] Even Chinese textile mills are establishing plants in South Carolina.[130] In the same vein, American corporations are going overseas to be close to their customers in addition to cheap labor and tax benefits provided in those countries.

The bottom line is that jobs can only be created if the United States has a growing economy, which only happens if there are stable and consistent government policies with a clear vision for the future that positively affect middle-class Americans (more about it in Chapter Nineteen). Even the definition of major economic indications, such as unemployment rate, needs a transformation to accurately reflect the new reality.

Unemployment Rate

One of the challenges in seeing how the middle class, and the country as a whole, is faring is that the definitions of some of the headline statistics (in the media), like the unemployment rate may be dated and may not account for structural changes taking place in the labor market. One would assume that the labor participation rate would be close to 100 percent if there are so many jobs available; instead it stood at 81 percent, with 2.9 million workers unengaged, in January 2015.[131]

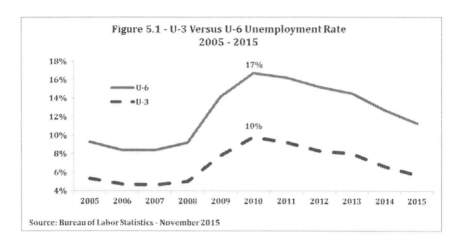

Figure 5.1 - U-3 Versus U-6 Unemployment Rate 2005 - 2015

Source: Bureau of Labor Statistics - November 2015

There is an official unemployment index published every month by the United States Bureau of Labor Statistics called U-3, and it defines the total number of people who are unemployed, as a percentage of the civilian labor force. There is, however, a U-6, a broader measure of unemployment that includes discouraged workers, under-employed, and part-time workers. There is a big disconnect between the two indexes that highlights the difference between the official unemployment and what middle-class people feel as shown by Fig. 5.1.[132]

One possible cause of the discrepancy could be a demographic shift where some people have left the labor force permanently.[133] It could also be that people are engaged in new professions that may not be included in the employment statistics, such as Uber drivers or Airbnb operators.

The good news is that both rates are coming down from a peak of 2010 as the criteria are further refined to reflect the alternative forms of employment such as the rise of the on-demand and sharing economy.

On-Demand Economy

On-demand economy refers to freelance workers who work for themselves and contract with large and small companies as independent contractors. It is also known as the 1099 economy referring to tax form 1099 needed to report their income to the IRS. Independent contractors and freelancers have been in existence for a long time. However, with the rise of smartphone applications, it has taken a whole new dimension and is becoming a significant portion of the economy. It has opened up new opportunities for middle-class workers. Some of the apps enable people to order groceries, home cooked meal deliveries, in-home beauty services, and on demand nurses.

Uber is the new phenomena, which started in 2009 and serves 300 cities across 58 countries. It has developed a mobile app for smartphones that consumers use to get a ride from an Uber driver. Anybody can apply for a job as an Uber driver. Upon meeting their criteria, drivers are approved to download the Uber app and are ready to start. They are alerted to any jobs for hire in their area, and they can choose whether to take it or not. In addition, Postmates is transforming the way local goods move around the city enabling anyone to get any product delivered in less than one hour. It is a platform that connects customers with local couriers who can deliver anything sold in any store or restaurant.[134]

Amazon started out as an online bookseller and now sells everything and anything they can and delivers to homes as early as the next day. Their goal is to deliver goods within 30 minutes to meet the needs of consumers and businesses wanting things NOW. They are even experimenting with drones to serve as a delivery vehicle from their local warehouses across the country for small packages. Small packages weighing less than five pounds make up 86 percent of the items they deliver.[135] Then there is a sharing economy where resources are pooled together to produce goods or provide services more efficiently at a lower cost.

Sharing Economy

The sharing economy allows middle-class people and businesses to earn money by renting their homes, car rides, production facilities, or computer data centers. The goal is to reduce waste in all areas. Smartphones and mobile apps are at the center of this economic change, allowing anybody to start a business by developing an app or utilizing one to satisfy consumer demand. According to a survey by PricewaterhouseCoopers, 86 percent of United States adults agree that these kinds of innovations make life more affordable, and 83 percent agree that they make life more convenient and efficient.[136]

Airbnb competes with hotels and motels by pairing up tourists/visitors and local hosts with homes and condos to rent out rooms for one night to a longer duration at significantly lower rates. It was valued at $24 billion as of June 2015—a larger market valuation than the hotel giant Marriott International. It gives middle-class people a chance to maximize their income by renting their extra space.

Businesses, large and small, are employing a distributed production model instead of owning their manufacturing facilities. Traditionally, manufacturers used to set-up their own factories to produce goods and employ dedicated staff and spent years in setting up their infrastructure. In a distributed production model, companies leverage a number of partners, commonly known as contract manufacturers. Contract manufacturers own the equipment and facilities and produce goods for a number of businesses, some of whom even compete with each other. This way a production facility can produce a multitude of products instead of one or two made by and owned by one company. As a result they can share the expenses accordingly in producing those goods. This model allows companies to save money and thus hire more workers. Moreover, it makes manufacturing jobs more secure since a factory doesn't depend on a single company for its income.

In this sharing model, businesses do not have to worry about spending years getting building permits and investing money in building a new manufacturing facility or finding talents to produce the goods. Contract manufacturers maximize the use of their assets by offering their equipment, human capital and facilities to a number of customers to maximize their use. Similarly, they use outside transportation companies instead of their own trucks to distribute their products through mega distribution centers managed by logistic companies who also distribute other companies' products.

Cloud computing offers computing and storage solutions to large and small businesses as well individuals through data centers. Data centers share and allocate resources to their customers across the globe. For example, when the United States is asleep and China is up, a data center can allocate their computing and storage capacity to China to meet their peak demand and revert back when the usage is heavy in the United States and vice versa. One can argue that this model has been instrumental in the rise of so many Silicon Valley companies that don't need to build their own manufacturing, computing or storage infrastructure in order to grow or even start a business.

The sharing economy primarily relies on mobile technology to match supply and demand of goods and human capital, and that flexibility is transforming society, creating new opportunities to those who are working and those who are forgoing traditional nine-to-five jobs.

Flexible Workforce

There are 53 million Americans who are doing freelance work, or about one in three persons in the workforce, and they contribute $715 billion in earnings to the national economy.[137] Uncertainty in the market caused by the 2008 recession and the financial crisis had a lasting impact on the corporate hiring practices. In today's economy, companies want the flexibility of increasing and decreasing their workforce quickly based on the supply and demand for their products. At the same time, qualified workers, especially the younger generation, are also demanding flexibility at work. They want their personal freedom and a working arrangement that meets their needs based on their performance at work. This pay for performance model is mutually beneficial to both employers and the talents.

With 5.5 million jobs unfilled and ever-changing business and staffing needs, businesses depend on temporary help to meet the short-term uptick in demand or when a staff calls in sick or cannot make it to work for a short period. Having that leverage, qualified workers are not only asking for better pay, they are placing flexible hours as one of the top perks they would like to have instead of traditional benefits.[138]

Wonolo, a new startup in 2013 with the support of Coca-Cola, offers the platform to find workers on demand. Wonolo is shorthand for "Work-Now-Locally."[139] Their vision is "for people to find work, whenever, wherever and for whomever they want and for companies to fill immediate needs with great people doing a great job, all at a moment's notice."[140]

Having a college degree helps get higher paying jobs. However, rising college costs are keeping a significant number of lower and middle-class students from earning a degree. In many cases, these emerging models value hard work and imagination over, and even to the exclusion of, degrees. They are empowering the lower and middle-class American workers to increase their earnings and put their personal priorities first while increasing national productivity.

In addition to the stagnant income side of the equation for the middle class, the most devastating part is the expense side, such as healthcare and education costs, student debt, and taxes that have grown at a much higher rate than inflation.

Six

United States Healthcare System

The nation is in the midst of a transformation of the United States health-care delivery system where power is transferring away from the providers and toward the patients, the consumers. The net result will be the reduction in healthcare cost as incentives are transferred to the patients due to higher deductibles and rewards offered by employers for maintaining good health. Americans spend more money on entertainment and car maintenance/ownership[141] than healthcare (not including health insurance).[142] Moreover, the American people have largely relinquished pricing power to healthcare providers who charge whatever they deem reasonable with little accountability. The next generation of healthcare delivery models will add value for consumers and use technology efficiently, which will result in significant overall cost reduction—exactly what the middle class needs.

Women are a critical element of healthcare delivery system. Women make 80 percent of a family's healthcare decisions, the main portion being the health insurance selection. Women also have a greater influence on healthcare decisions by exchanging ideas and referrals through social media; 70 percent of them use social media where they influence healthcare decisions in their online communities.[143] As a result, women can bring a consumer's perspective to the system as it transforms.

Women are 74 percent of workforce in the healthcare field. The rise of a healthcare retailing model calls for physician assistants (PA) and nurse practitioners to manage non-critical medical needs at an affordable cost. Nearly 75 percent of physician assistants, 91.5 percent of nurse practitioners and almost 72 percent of the medical and health service managers are women.[144]

Some of the prominent women leaders in the field are Sylvia Mathews Burwell, the Health and Human Services Secretary under President Obama; Dr. Susan Desmond-Hellmann, the CEO of the Bill and Melinda Gates Foundation, which is focused on global health; and Sue Siegel, the CEO of GE Healthymagination, which harnesses innovation and partnership to improve the quality, access, and affordability of healthcare.

The rise of women's voice in healthcare is needed and will lead to better quality and lower costs for the middle class. The health industry has historically been a male dominated field, especially in the role of doctors and the field of gender-based research. The one-size-fits-all model ignores the healthcare needs of women, despite the obvious differences between male and female bodies. For example, heart disease develops seven to ten years later in women than men, yet it is still the highest cause of death in women. The risk of heart disease in women is often underestimated due to the misconception that women are protected against such diseases.[145] Depression affects women at twice the rate it affects men. Two-thirds of the 5.1 million people with Alzheimer's are women.[146] Osteoarthritis affects more women than men.[147] Ever increasing numbers of women entering the medical field will broaden gender-based research, which will result in more effective treatment for women, saving lives and money.

Even though gender-based research still lags behind, there is progress: the nation now has offices of women's health in several states and most U.S. Department of Health and Human Services (HHS) agencies. There are also twenty-one centers of excellence in women's health and more than $3.8 billion allotted by the National Institutes of Health (NIH) for women's health research.[148] Furthermore, the number of women who are physician-researchers leading studies in prestigious medical journals has increased almost five fold from 1970 to 2004. However, the number and prominence of women physician-researchers still lags behind men.[149] As the number of women graduates in public health and medicine grows so will the number of women researchers that will bring better solutions for women's healthcare.

Legacy Healthcare Delivery Model

Doctors and hospitals are businesses after all, and account for 50 percent of healthcare cost, along with drug companies at 9 percent and health insurance companies at 6 percent in 2013 (Fig. 6.1).[150]

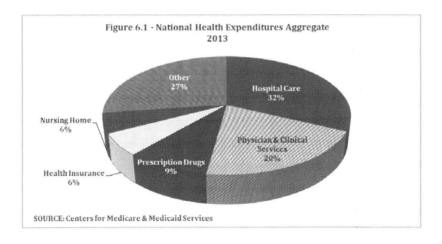

Figure 6.1 - National Health Expenditures Aggregate 2013

Hospital Care 32%

Physician & Clinical Services 20%

Prescription Drugs 9%

Health Insurance 6%

Nursing Home 6%

Other 27%

SOURCE: Centers for Medicare & Medicaid Services

The United States healthcare delivery model is inefficient. Patients do not have full pricing information upfront but are expected to pay whatever is charged regardless of the quality of the care or service. There is very little accountability or transparency. In addition, healthcare service providers have incentives to increase the healthcare costs since they are paid for the procedure regardless of the outcome. Therefore, the more procedures there are; the more money for them. At the same time, consumers have no incentive to challenge the costs since most of the cost is paid by a third party, so they do not even get to see the bill. The current healthcare delivery system rewards for procedures and care, but there is very little incentive for prevention. "When someone has a congestive heart failure, we pay a lot of money to the care system for that. But if that same organization prevents the failure by intervening with the patient, helping the patient, making sure they get the right medication, they don't get paid for it," said George Halvorson, former CEO of Kaiser Permanente, one of the biggest nonprofit hospital systems.[151]

Two factors that are keeping the healthcare system from responding to free market principles are the fee-for-service model, which requires patients to pay for services regardless of the quality or value of the care, and the third-party payer model where somebody else pays the medical bills. Medicare (discussed in Chapter Eleven) and similar programs keep consumers from demanding accountability for cost, simply because consumers aren't paying attention when they aren't the ones paying the bills. The same goes for employer-provided healthcare. Since employees' contributions are deducted from their paychecks, they are less likely to notice and question the cost.

The notion that the healthcare industry is different than other private industries and exists outside the free market model discounts the fact that healthcare is a $2.6 trillion business and almost all of the key stakeholders are for-profit corporations. Most hospitals, which make up the largest part of the healthcare cost, are owned by large corporations. Even the nonprofit ones operate more like for-profit hospitals with high salaries for their executives, but they do not pay any taxes. Drug companies and health insurance providers also make billions in profits, and their executives make millions, which ends up reflecting in higher costs for their products and services.

The missing part in all this is the accountability and transparency that make a system efficient. While it's certainly a complex situation, patients should have the right to question and hold providers accountable about the quality or cost of service and have justification of the bill, or access to their own medical records.

Out-of-pocket payments for medical expenses fell from 47 percent of total healthcare spending in 1960 to a record low of 12 percent in 2008 (Fig 6.2). At the same time, there was dramatic growth in the public funding of healthcare, reaching a record high of 47.3 percent in 2008—up from 24.5 percent in 1960.[152]

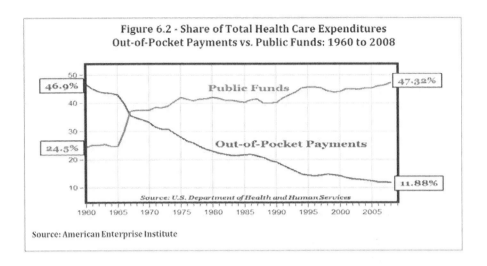

Figure 6.2 - Share of Total Health Care Expenditures
Out-of-Pocket Payments vs. Public Funds: 1960 to 2008

Source: American Enterprise Institute

According to an American Enterprise Institute report, "As consumers have relied more and more on employer-provided healthcare and government programs such as Medicare and Medicaid, they have become less and less conscious of healthcare costs because they have been increasingly spending somebody else's money, and not their own. Just imagine what would happen over time to the cost of food, clothing, or automobiles if consumers paid only 12 percent of the total bill, with the other 88 percent is paid by employers or the government, and it's easy to understand why healthcare spending goes up year after year."[153]

Employer provided healthcare programs exacerbate ever-rising healthcare costs. If a person makes $50,000 and their employer's health plan is worth $12,000, it is not taxable to the employee and is tax deductible to the employer. However, if the employee decides to forgo that benefit and asks for it to be paid as a salary, it will increase his or her taxable income to $62,000. So now he or she will pay taxes on $62,000 as well as an additional 7.65 percent in payroll tax. Therefore, employees are forced to accept the plan offered rather than take the money, take the responsibility of their healthcare choices, and perhaps buy a much cheaper catastrophic health insurance plan and have more purchasing power.

Even though employers and health insurers have been trying to move the consumers to low cost plans and healthier behavior to cut their cost, consumers have little incentive to change since the employer pays much of their premiums. Some of the large employers are trying to provide financial incentives by allowing employees to keep some of the money they don't spend on healthcare from the allocated budget. In addition, employees get punished or rewarded depending on how well they manage their health. IBM offers $300 a year in rebates to its employees for participating in an exercise and nutrition program, which saved them $190 million in healthcare costs between 2005 and 2007. Eighty percent of the employers planned to financially reward employees for healthy behavior.[154]

The emerging healthcare model could follow a rewards program offered by credit card companies except in reverse: consumers would be rewarded points for the money they save on doctor visits and procedures by shopping for their medical care—such as Smart Patient Credits (SPC), discussed in Chapter Ten.

There are, of course, other factors besides price to consider, such as quality of care. Also, there are certain situations, such as emergency treatments, where it may be impossible to know the costs up front, but there should still be more transparency and accountability than under the current system. However, for doctor's office visits, flu shots, and standard, well-established elective procedures such as gall bladder removal and knee replacements, a price or at least a price range should be set and clearly stated, as it would be in any business.

Transparency & Accountability

Affordable wellness can be accessible to every American. However, it would require personal responsibility as well as accountability and transparency by the healthcare providers. Under the current healthcare delivery model, there is neither. Unfortunately, Americans have been conditioned to think that healthcare costs will continue to rise at a higher than inflation rate and in comparison to any other goods and services. In May 2009, this phenomenon was underscored in a meeting of healthcare lobbyists with President Obama where they pledge to cut the *rate of growth* of healthcare spending by 1.5 percentage points each year.[155] Therefore, the goal of all stakeholders, including the government and the healthcare providers, is simply to slow the growth of healthcare costs instead of reducing the cost of healthcare so that it is affordable for every American.

Employers are increasingly shifting toward health insurance plans with high co-pay and higher out of pocket expense to keep their cost down. As a result, the percentage of covered workers with annual deductibles of $1,000 or more has increased from 10 percent in 2006 to 46 percent by 2015.[156]

As this cost shifting continues, consumers will demand accountability and transparency from their healthcare providers. For example, Joe Weinzettle of Florida was taken to an emergency room after an accident. He had no injuries or complaint of any pain and did not feel that he needed medical attention, but authorities insisted he seek medical attention. He ended up with a $45,000 bill for his three-hour stay at the hospital. He started a petition on Change.org calling on the Federal Trade Commission (FTC) to investigate that hospital for its pricing practices. He had 116,906 supporters for that petition as of October 2015.[157] The petition was received by the FTC and is pending further action.

Individuals who have coverage through employers do not have the financial incentives to maintain their health as they do with a house or car. They pay a very small percentage of their total bills. When a problem occurs they go to the doctor or emergency room and there is no financial incentive to prevent the problem. Those who do not have any health insurance coverage, due to lack of funds or lack of eligibility for plans, wait until an emergency occurs and then rush to the hospital knowing that they will be taken care of by law. Healthcare providers do not have any incentive to be transparent about their pricing or billing, knowing that the consumer does not ask for it, aside from copay amounts.

More than 62 percent of the personal bankruptcies can be attributed to medical bills, according to a national study by the American Journal of Medicine.[158] The biggest hurdles that patients face is that they have no recourse in challenging the bill whether it is from the hospitals or doctors, two major categories of the United States healthcare cost equation. The bills are impossible to comprehend since they are full of acronyms and codes that even hospital administrators don't always know. Patients can try to challenge the bill but may end up facing the collection agency.

Changes are already beginning. In 2014, the credit scoring firm, Fair Isaac Corporation, widely known for credit score FICO, started to reduce the impact of medical debt on its scoring model. Then in March 2015, three major credit bureaus, Equifax, Experian, and Transunion announced changes in the way they report medical debt as part of the settlement with the New Attorney General's office. This will empower the consumers, most of them middle class, to challenge the validity of charges and the pricing of services without destroying their credit ratings.

Healthcare Redefined

Skyrocketing healthcare costs, the Affordable Care Act (ACA), increasing out of pocket expenses and technology will serve as an impetus to transform the healthcare delivery system in the coming decades. A new paradigm shift is underway where decentralized and sharing models with emphasis on personal accountability is evolving that will make healthcare accessible and affordable for all.

The call for enhancing value and outcomes will be amplified as the healthcare cost burden transfers directly to consumers. An increased focus on prevention will change life for the middle class and beyond—up to 360,000 deaths per year can be prevented.[159] Prevention includes primary, secondary, and tertiary preventive services: primary interventions to prevent the disease before it occurs by altering unhealthy behavior, secondary intervention to reduce the impact of a disease by reacting quickly to slow down its progression through regular exams, screening, exercise, etc., and tertiary prevention to reduce further worsening of a chronic condition and improve quality of life.

The decentralized distribution model will transfer healthcare delivery from doctors and hospitals as main providers for routine and nonemergency services to local pharmacies, Costco, Walmart, and more, creating new types of outpatient storefronts that have not been imagined yet.

The sharing model will facilitate a continued push for medical records to become digital so that they are accessible by all providers in real-time, which will eliminate duplicate records and tests. Moreover, patients will be able to carry all of their medical records on a flash drive attached to their keychain for quick access at any time just like their driver's license or an ID card.

As the Consumer-Driven Healthcare (CDHC)[160] model gains acceptance, healthcare will be more affordable. American consumers will have more incentive to take care of their health and monitor costs since they will pay more out-of-pocket expenses. There will always be a number of Americans who cannot afford the insurance or pay for services from their pocket, and that is where the government has a role to play as discussed in Chapter Eleven.

The following chapters will look at innovations and advances that are bringing changes to each sector of the healthcare industry. There are five major components that make up most of the healthcare costs: hospitals, doctors, drugs, health insurance companies, and Medicare.

Seven

HOSPITALS

The factors that adversely impact the healthcare system as a whole specifically affect hospitals since they represent the largest (32 percent) share of the United States healthcare cost (Fig. 6.1). This chapter will examine the innovations that are happening or have been envisioned that can change the way hospitals work, benefiting the middle class. Technology, the leadership of women, and more transparency can change health outcomes for patients. More importantly, change will drastically reduce hospital costs.

Boston has an unprecedented number of women leading their major hospitals reflecting their growing role in the healthcare field. Seventy percent of the healthcare service managers are women,[161] and as these women rise in the ranks, more and more will reach the executive level. However, women are still underrepresented on hospital boards. Nationwide, only 15 percent of all medical chiefs are women.[162]

Some of the prominent women leaders in hospitals are Sister Carol Keehan, the president and CEO of the Catholic Health Association with more than 600 member hospitals; Donna Lynne who is responsible for Kaiser Permanente's Pacific Northwest and Hawaii regions in addition to leading its Colorado health plan; Patricia Maryland, President of Healthcare Operations and Chief Operating Officer at Ascension Health, one of the largest nonprofit health systems that provides nearly $2 billion in care to people living in poverty through 129 hospitals and thirty senior living facilities; and Judy Faulkner, founder and CEO of Epic Systems that sells healthcare software to hospitals such as Cleveland Clinic, Mayo Clinic, and Johns Hopkins.

Technological innovations will significantly reduce the need for hospitals as new venues, such as local drug and grocery stores or big box stores, enter the market. Physician assistants, experienced nurses, or nurse practitioners will manage these locations for routine medical conditions that do not require a physician's presence—at a significantly lower cost, possibly by half. This will allow people with routine but not life-threatening conditions from ending up in the emergency room on evenings and weekends when their family physician isn't available.

According to George Halvorson, "1.7 million Americans go to the hospital every year and get infections that they did not have the day they went to the hospital."[163] Nearly 100,000 people die each year from infections acquired while in the hospital, according to the Center for Disease Control and Prevention.[164] As the dependence on hospitals declines, a significant number of those lives would be saved partly due to the reduced duration of patients' stay there.

Amazing technological developments are in progress, which will reduce the hospital costs, provide quality care to patients, and cut down on infections acquired during the hospital stay. The future hospital rooms will be integrated with all monitoring devices. The high-tech bed will weigh the patient, shift a patient to eliminate the risk of bedsores or alert the nurse should the bed get wet. All that data will be fed directly to the patient's electronic health record in real-time. This will provide a warning time for doctors and nurses to respond to an impending heart attack or respiratory failure or any other kind of life threatening condition. All the while, these technologies will save time for nurses and doctors from routine functions so that they can spend more time with the patient. The more time they spend at the bedside, the less likely patients are to fall or suffer an infection. In addition, medical errors decline, which would be critical for hospital reimbursement as the society moves to a pay-for-performance based model.[165]

Robots will roam the corridors and hallways of most hospitals—twenty-five already do at UCSF Medical Center at Mission Bay, California since 2015.[166] They deliver meals, linens, and even medications to patients and pick up trash. This saves money and minimizes the transmission of infection by reducing human-to-human contact.

Hospital technology will also become increasingly portable. IBM's first gigabyte hard drive weighed 550 pounds and cost $40,000 in the 1980s. By 2015, three terabyte hard drives with 3,000 times more storage cost around $100 and weighed about three pounds. The same trend in technology will miniaturize the X-ray, CAT scan, MRI, and other bulky machines so that they are portable and can be taken to patient's room rather than bringing patients to a special screening room. Today bulky MRI and CAT scan machines are normally housed in the basement of a hospital. All day and night patients are hauled from their room and taken down to the screening room for the scan. It adds hardship to the patient, adds labor and administrative costs in moving the patient, and increases the chances of infection by having more people involved in the activity.

Less invasive devices like wireless pacemakers that could be implanted without a surgery would reduce surgical costs for the 200,000 people every year who need a pacemaker.[167] Remote monitoring of patients 24/7 using high tech devices can reduce the hospital readmission rate after surgery by 60 to 70 percent. It can also reduce the average number of days a patient needs to stay in the hospital.[168]

Ingestible sensors that look like pills and are powered by stomach fluid can transmit information about the activities in a patient's body to the medical service provider in real-time through wireless devices, such as a smartphone. It can even replace an annual physical visit to the doctor's office since the pill collects the data and can transmit that to all authorized professionals.[169]

It is not farfetched to imagine that most of the noncritical care will be managed at home instead of a hospital. Doctors will not only be able to monitor a patient's vital signs but visit the patient virtually via the Internet from their office or home and prescribe medicine if needed. This could potentially reduce the 400,000 deaths per year that occur due to preventable harm to patients during their stay in the hospital according to a report in the Journal of Patient Safety.[170]

A combination of the research and innovative concepts will help further reduce the cost of healthcare.

Predictive Versus Reactive Medicine

The Center for Disease Control (CDC) publishes an Influenza Surveillance Report every week for the previous week's data. However, Google, based on the people's search activities developed a model where it could predict the spread of the flu based on the search activity. They compared the fifty million most common search terms used by Americans and compared that with the CDC data for the flu for five years and developed a mathematical model. It ended up with forty-five search terms that had a strong correlation between their prediction and the CDC data.[171] Some of the top queries most correlated with CDC data were Influenza Complication, Cold/Flu Remedy, General Influenza Symptoms, and Term for Influenza.[172] Unlike the CDC data that is delayed by a week, the Google model can offer real-time data that could provide advanced warnings to hospitals and healthcare professionals so that they can be better prepared for any onslaught of cases.

Predictive analytics utilizes big data through queries on search engines, such as typing in symptoms, along with discussion forums where people with the same condition share information. Big data is described as a very large volume of data that could be analyzed to identify trends, patterns or any correlations that could be used for better decision making. The hospitals then can use this data to more effectively treat patients. Hospitals are already able to use big data to personalize medical care for patients by analyzing the effectiveness of their protocol. Since 2013, Mercy Hospital's (Cincinnati) diabetic patients showed an 11 percent improvement through this method.[173]

Organs on Demand

3-D technology will be a game-changer in the healthcare field. It opens the door for personalized medicine that will solve the unique need of patients whether it is implants for knees or teeth, or reproducing an organ. Scientists at Carnegie Mellon University are working on reproducing organs and showing some success toward the reproduction of a human heart. It could save 4,000 Americans who are on a waiting list to receive a heart transplant[174] and the healthcare costs associated with that waiting. This is just the beginning. Producing organs through 3-D printing will significantly reduce the cost of caring for people with failing organs, and it will increase their lifespan and quality of life.

As 3-D printing continues to develop, it will become more common for hospitals to have 3-D printers to produce limbs and other prosthetics on demand. Bio-printing—mixing human cells, water, and biocompatible material to produce living human tissues—could allow hospitals to generate skin or organs. Scientists have already produced blood vessels and bone tissues.[175] At a cost of $10,000 for a 3-D printer, savings for such procedures would be incalculable. Producing custom organs on demand would save the lives of hundreds of thousands of people who die waiting for organ donations.[176]

Cash Payment Discounts

One more positive sign toward the healthcare cost reduction is cash payments by the patients as they realize that they can pay much less to hospitals by paying cash. It appears that most of the medical care costs are artificially inflated given that most of the hospitals and doctors offer discounts, sometimes steep discounts for a cash payment. According to a *Los Angeles Times* article, one hospital in Los Angeles charged $6,707 for a CT scan of the abdomen and pelvis after a colon surgery, and the patient's share was $2,336. However, the patient would have paid only $1,054 if she had paid cash and not used her insurance.[177]

This growing awareness of cash pricing would accelerate the trends toward more transparency as well as lower overall cost to the society. One would argue that cash pricing would hurt those who could not afford to pay which is true. However, in light of the trend of higher deductibles and as the example above showed they would end up paying more even by having insurance. The other option for them is to go to the emergency room, which would be the case with or without the cash payment model.

The second largest component of United States healthcare expenses are doctors and clinical services, which are going through a major transformation with the help of technology and market conditions.

Eight

Technology will transform the healthcare model from doctor-and hospital-centric to patient-centric. Patients will have quick access to doctors—some call it the Uberization of healthcare. Doctors will be proactive and their offices will become paperless and efficient, and they'll communicate with health insurance companies in real-time. Local drug stores, smartphone apps and other new venues will compete for noncritical care, so patients will pay a much lower fixed price per visit and will be in and out in a few minutes instead of waiting for hours in a doctor's office or in an emergency room. All of these advances will make healthcare more affordable and accessible to middle-class Americans.

Women are leading the way in emerging healthcare delivery models; more than 50 percent of women doctors are in pediatrics or primary care. Their role will continue to grow as more women earned medical degrees—48 percent in 2011 compared to only 5 percent in 1952. Moreover, 91 percent of the registered nurses were women as of 2011.[178] As a result, women will play a critical role as the decentralized healthcare delivery model takes hold.

Doctor's Office of the Future

The legacy model of medical practice is giving way to a very efficient model that will take full advantage of technology and patients' adaptation of it.

It will allow doctors to spend more time with the patient, which will help minimize the misdiagnosis. Wearable technology, sharing of digital records and very small waiting rooms are a few of the venues and concepts that are in play.

Mobile and wearable devices are empowering Americans to manage their own health and transfer healthcare data from patient to doctors in real-time. They track a person's heart rate, calories burned rate, sleep pattern, and other data to help maintain better health. Directly connecting these devices to doctors who remotely monitor the patient will allow them to identify potential conditions even before the patient notices the problem. This will result in more proactive responses that could potentially save lives, let alone the cost of going to an emergency room. A good example can be a doctor who is treating a number of patients for blood pressure with a certain medicine. Wearable technology will allow the doctor to track the effectiveness of that medicine on each patient. Through big data and by aggregating and analyzing the data from this group, a doctor can determine its effectiveness based on different demographics and may customize the dosage or change the medicine for better results. Soon doctors could be chasing patients instead of patients looking for doctors.[179]

When they visit a doctor's office, patients will be able to go straight to the examination room by alerting the doctor via their smartphone app that they are in the parking lot and will be in the doctor's office in five minutes, practically eliminating the waiting room. The visit will start as you walk in since all the data including images will be on the doctor's computer and all the pre-visit questions were submitted electronically a day before the visit so that the doctor has the patient's updated information as well as a perspective on his or her medical condition. This would lessen or almost eliminate the wait time. Similarly, if a doctor is running late or has an emergency, he or she can inform the patients about the delay or reschedule the appointment saving them the time and cost of travel to doctor's office.

Up to 35 percent of visits will happen virtually, eliminating the need for office visits for routine checkups.[180] The proliferation of smartphone apps is empowering consumers to manage and monitor their health at very little or no cost to them. They can monitor their blood pressure, heart rate and may remotely transmit electrical heart signals (ECG) to their doctors. Developments in wireless and sensor technology will accelerate the cost reduction of overall healthcare expenditure—doctors and clinical services represent 20 percent of the overall healthcare cost (Fig. 6.1).

Some doctors are working to minimize their dependence on Medicare, Medicaid, and insurance-supported patients so that they can spend more time with patients by saving money in overhead due to less paper work. That has led to some more efficient models to improve the quality of care such as Concierge Medicine.

Concierge Medicine

A new breed of doctors are turning to a cash only model; they are tired of spending one-third of their workday[181] dealing with insurance companies and Medicare, which also requires a larger support staff. The main benefit of this model is that doctors get to spend more time with their patients, at a lower cost to themselves, and simply practice medicine, which is the main reason why they went to medical school. Patients get more time and personal attention from their doctors and the flexibility to switch providers anytime they are unhappy with the service or the quality of care, without waiting for approval from insurance.

Even though this model is not for everybody, it is another alternative for some who are healthy, want to maintain good health, have higher deductibles, or do not want to buy expensive health insurance. These consumers may just buy a catastrophic health insurance to avoid the ACA penalty and cover themselves from large unforeseen illnesses or injuries.

Concierge medicine, like buying a warranty on a car or an appliance, would provide unlimited access and more time with doctors for an annual, monthly fixed, or per visit fee that ranges from $50 to $150.[182] However, the patient will be responsible for any tests and can shop for better pricing in the marketplace. The doctors can limit the number of patients under this model and still make about the same amount of money as they would in a traditional model by saving one-third of their time from paperwork and overhead expense.

For those who cannot afford or choose not to buy concierge service, local pharmacies, supermarkets, or big box stores are starting to offer noncritical medical service at affordable prices.

Rise of Retail Healthcare

Traditionally, a person calls the primary care doctor's office for any health issue or goes to a hospital in case of an emergency. Calling the doctor's office normally means waiting on the phone or leaving a message and waiting for a call back. Then it may be several days or weeks before a patient gets to see the doctor.

Higher deductibles and better technology are creating opportunities for drug stores like CVS and Walgreens and big box places like Costco, Walmart, and others to expand their customer and revenue base. These places are already where customers come to buy healthcare products so healthcare services are a natural extension of their offerings. They can hire a physician's assistant or nurse practitioners to easily manage the basic medical tasks at a much lower compensation than a physician. Therefore, they can afford to offer those savings to the consumer in lower fixed cost for tests or visits. In 2015, Walgreens had 400 of these clinics and CVS had 1,000.

As employers and health insurance companies offer incentives toward prevention more consumers would move toward retail outpatient facilities for routine checkups and minor medical conditions. One of the main reasons would be the reduced cost of consultations and tests compared to doctor's offices or hospitals. The second would be the convenience of walking into a retail facility and getting taken care of within minutes instead of waiting for hours, days, or even weeks. Millions of websites provide a vast amount of information on every aspect of healthcare that helps consumers choose their providers and ask the pertinent questions.

Doctors on Demand

It seems as if the United States is going back in time to when doctors used to make house calls. In the 1930s, physician house calls accounted for 40 percent of medical visits, which dropped to 1 percent by 1980. Now they are back at $39 per virtual visit with a waiting time of five minutes or less, offered by VirtualHealthNow.[183] Doctor on Demand, cofounded by TV personality Dr. Phil, is another such service where one can visit a doctor via smartphone or desktop for $40 and higher based on the session time without leaving home.

Those patients who can't get out of bed or do not want to spend hours waiting in the emergency room can have a physical visit by a doctor. A doctor will visit a person's home within one to two hours at a cost of $50 to $200 depending on the service and the location. Heal, a company founded by Dr. Renee Dua, offers physical visits by a doctor seven days a week in some cities in California at a flat fee of $99.

These services are significantly cheaper than going to a doctor's office, especially with the cost of time lost in waiting.

Cost of Patient Time for Doctor Visits

Public discourse on healthcare costs often focuses on how much or how little doctors and hospitals are paid by insurance companies or Medicare. However, there is no discussion on the cost to the patient in waiting for a consultation with a doctor.

The current model serves patients at the convenience of the doctors and the hospitals. This model places undue burden on the lower and middle-class patients because of lost income while waiting at the doctor's office—not to mention transportation costs for a visit.

Since doctors get paid by the insurance companies or a third party, they have every incentive to schedule as many patients as they can to maximize their income without worrying about the patients' time. Patients are accustomed to wait without complaining since they pay very little out of pocket toward their medical bills.

Doctor's visits cost patients $43 in lost time per medical visit, according to a study by the Harvard Medical School. The study estimates that a typical visit to a doctor consumes almost two hours of a patient's time—thirty-seven minutes in travel, sixty-four minutes waiting for care or filling out forms, and only twenty minutes face to face with the physician. Americans spent 1.1 billion hours, or $52 billion, per year to obtain medical care for themselves or others—in addition to the doctors' or hospital charges.[184] These costs have no economic justification and place undue burden on the lower and middle class. Virtual visits and even house calls minimize this costly burden.

Defensive Medicine

Defensive medicine burdens the American healthcare system with very little or no benefit to the patient. Defensive Medicine describes the practice of physicians ordering tests and procedures or taking other costly steps that may not be necessary in order to protect themselves from any liability. A good example can be having a CT scan done when a routine X-ray at a significantly lower cost would have been sufficient. Six out of ten doctors place defensive medicine on the top of the list that contributes to rising healthcare costs.[185]

It is unrealistic to expect the doctors to take personal financial responsibility for an honest human error. The system puts undue burden on doctors to provide error-free service, which is humanly impossible. Therefore, the practice of defensive medicine continues at a cost of $650 billion to $850 billion per year.[186]

Nine

DRUGS

Incredible opportunities are on the horizon due to research and the development of new technologies where each patient will have a customized drug treatment and protocols that can be monitored in real-time and adjusted according to the patient's need. 3-D printing may allow a pharmacist to create a single customized pill with the ingredients of all the medications a particular patient needs. Someday there may simply be a skin patch that releases the required medicine at a precise time for months at a time. All of these innovations would make life better and save money for the middle class.

Entrepreneur women are pioneering some of the blood and gene testing innovations that could save millions of lives by proactively addressing the disease instead of reacting to a symptom. They are challenging the centralized $54 billion dollar industry[187] by offering tests that can be cost effective and efficient for patients and doctors. Some of those tests offer results in real-time instead of days saving critical time to respond to an illness immediately.

Amy Winslow is the President and CEO of Magellan Diagnostics,[188] a medical device company that provides point-of-care systems. Applying a novel technique called Anodic Stripping Voltammetry (ASV), the company designed and built a bench top analyzer for the measurement of lead in blood. The company also provides portable instruments that can be used outside the traditional central labs. As laboratory testing devices become smaller, blood and glucose tests among others could be done at the point of care, either at the patient's home or nearby drug stores.

Nanobiosym, another women-led company founded by Dr. Anita Goel, is focused on decentralizing, personalizing, and mobilizing diagnostics through its Gene-RADAR device. This device, the size of a smartphone, takes a human DNA sample from a drop of a blood or saliva and determines the nature of the disease in minutes. Nanobiosym's mission is to bring world-class healthcare to more than four billion people around the world who lack access to centralized diagnostics.[189] The device does not need running water, constant electricity, or highly trained personnel to test for AIDS, HIV, E. coli, and some types of cancer. Today it costs about $200 for an HIV test and takes two weeks to get results. However, through Gene-RADAR the same could be accomplished for about $2 in minutes—saving people valuable time especially in life-threatening conditions.

Anne Wojcicki, a former Wall Street executive, cofounded 23andMe, which provides personal genetic tests for under $200 using a person's saliva. The test shows whether a person carries genes related to certain illnesses that can be passed on to their children such as Parkinson's disease or cystic fibrosis. Their mission is to help people access, understand, and benefit from the human genome to make proactive decisions about their health and life. This will lead to personalized medicine since each human is different.

Jennifer Doudna, a biochemist at the University of California, Berkeley, who helped make gene editing as simple as editing a word document, was recognized as one of the 100 most influential people in the world by Time magazine.[190] Her discovery led to an easy way to alter any organism's DNA and was awarded a $3 million Breakthrough Prize endowed by leading Internet entrepreneurs.

Personalized Medicine

Personalized medicine, or precision medicine, has the potential to reduce healthcare costs through prevention and efficient treatment for each person's unique condition, genetic makeup, and environment.

The Human Genome Project was completed in 2003, giving scientists a blueprint of the human body, in order to identify specific genes responsible for certain diseases. The human genome identifies an individual's predisposition to certain medical conditions and its response in fighting diseases. These findings can then generate a solution through individualized diagnostics and treatments including drug therapy. Personalizing in this way means a higher probability of a desired outcome and reduced probability of side effects.[191] Moreover, focusing on prevention through genetic understanding reduces healthcare costs. A study by the Mayo Clinic showed that hospitalization rate for heart patients were reduced by about 30 percent when genetic information was available to doctors prescribing the drugs.[192]

In the summer of 2015, the Food and Drug Administration (FDA) gave approval to a pharmaceutical company to develop patient-specific pills made by 3-D printers. Even though this is a few years away, this technology will allow pharmacists to make a custom medicine to each patient's needs and adjust the doses as needed. Combining several drugs into one will reduce the number of pills a patient takes all into one.[193]

Pharmacogenomics

Pharmacogenomics is a study of how genes affect a person's response to drugs. It combines the pharmacology and genomics that results in personalized medicine. Traditionally, drugs have been developed in a centralized mode as one-size-fits-all. No two human bodies are the same and their genetic composition varies. Therefore, doctors use trial and error hoping to find the right dosage, a very expensive way to manage drugs protocol. The study of genomics will pinpoint the most effective solution from the beginning, which will save time and money and the human suffering of trial and error.

The first human genome required $3 billion and thirteen years to sequence, which is down to $1,000 and twenty-four hours as of 2015.[194] Gene sequencing machines will be in every hospital just like MRI and ultrasound machines as their prices continue to come down. Once the abnormality of the gene is identified, the next challenge is to modify it through gene editing so that it behaves like a normal gene.

Gene Editing

Another revolutionary yet controversial treatment is known as gene editing. Editas is a startup led by CEO Katrine Bosley that is planning to start a clinical trial in 2017 to treat a rare form of blindness using CRISPR (Clustered Regularly Interspaced Short Palindromic Repeat), a ground breaking gene-editing technology that can edit a person's genes.[195] The CRISPR system is a collection of molecules that work together to edit, modify, or to correct DNA disorders that results in terminal illnesses.

Gene editing is a three-step process: 1) Identify the abnormal section of the DNA causing the disease, 2) Use a molecular scissor to remove that section, 3) Replace it with a normally functioning DNA sequence. This technique has the power to prevent Alzheimer's, cystic fibrosis, and other terminal diseases.

Distributed Medical Testing

Over the next decade, the healthcare system will transform from a reactive model to a preventive one. Point-of-care (POC) or decentralized medical testing is yet another factor in transforming the healthcare delivery model. It provides immediate, less expensive, and more accurate results as opposed to the current centralized testing model that takes days to get the results at a much higher cost. Cholesterol, blood pressure, glucose, urinalysis, or even cancer marker tests are being performed at local drug store clinics. This method can greatly reduce the chances of human errors and sample mix-ups.

Having more affordable medical testing avenues will encourage Americans to test more frequently as part of their routine healthcare maintenance program. The net result will be an earlier identification of any potential medical condition, which will allow time to respond proactively—saving pain, hardship, and costs.

High Cost of Bringing Drugs to the Market

While the POC accelerates the pace of understanding potential medical conditions, the cost of bringing innovative drugs to the market continues to increase. Disruptive innovations have leapfrogged the legacy medical testing from days to real-time results. However, it takes 10 to 12 years to bring a new drug to the market—an eternity in today's rapidly changing environment. At the same time, only half (11.8 percent) of the drugs that enter clinical testing get regulatory approval compared to the 1990s, according to a study by the Tufts Center for the Study of Drug Development. It costs $2.3 billion on average to bring a new drug to the market.[196] There are arguments that the drug companies exaggerate that cost to justify higher prices. This Tufts Center study was funded to a large extent by the pharmaceutical industry.[197] However, the fact remains that a decade to bring the drug to market is too long, and the cost in human lives that could be saved is too high to quantify.

Another factor leading to higher prices is that Congress required all drugs approved between 1938 and 1962 be reviewed for their safety and effectiveness and be approved again to make sure they meet today's standards.[198] The concern was that those drugs may not meet today's testing standards. However, that adds to the cost of the drugs. Drug companies, on the other hand, have used that act to increase the prices by up to fifty-fold without altering or improving the products.[199]

FDA's decade long approval process requires transformation in light of the fast paced innovative environment because some of the drugs may become obsolete during the process as personalized medicine takes hold. One of the solutions could be allowing the drugs approved in the developed countries to be sold in the United States without further lengthy approval. The cost savings could be in the billions. A second possible solution is allowing or modifying the definition of off-label use of drugs. FDA approved drugs give information about the approved dosage and how it should be given to treat the medical condition for which it was approved. However, doctors find that some of the approved drugs could be as effective for other conditions which will then be defined as off-label use of drugs. An example of off-labeling is when a drug is approved for treating one type of cancer but is then subsequently used to treat a different type of cancer.[200]

Pay-for-delay is another way drug costs remain high. A brand-name drug maker can pay its competitor to keep the lower priced generic drug off the market for a number of years. It is more profitable for brand-name drug makers to pay generic drug makers not to produce the generic version of their brand name drugs so that they can keep their drug prices higher. In 2013, the United States Supreme Court ruled that pay-for-delay arrangements violate the antitrust laws. This ruling will increase the availability of generic drugs. According to the FTC, without those arrangements, consumers will save $3.5 billion annually.[201]

Monopoly Power Disrupters

Brand-name drug makers appear to have all the power due to patent protection and its influence on the political deliberation and over patients' health. For example, it's illegal for individuals to import medical drugs to the United States even from Canada,[202] and Medicare cannot negotiate drug prices.[203]

As a result of these protective laws, brand name drug prices have been skyrocketing: Derma-Smoothe oil for eczema went from $46 in 2009 to $323 by 2015. Turing Pharmaceutical generated public outcry in September 2015 when it raised the price of Daraprim` (used to treat a parasitic disease that especially affects people with AIDS), from $13.50 to $750 per capsule. But Imprimis Pharmaceuticals, a specialty pharmaceutical company based in San Diego, responded by offering a compounded formulation as a low cost alternative to Daraprim` for as little as $1 per capsule and still be profitable.[204]

As the market shifts toward personalized medicine, the importance of compounding pharmacies grows. They are managed by a licensed pharmacist to meet the specific need of a patient by customizing drugs that may not be commercially available. These drugs are not FDA approved but their ingredients are, and federal and state authorities regulate the pharmacies.

Direct to Consumer Pharmaceutical Advertising (DTCPA)

The United States is the only country in the world besides New Zealand that allows direct-to-consumer advertising of prescription drugs, such as Viagra, Cialis, or cholesterol- reducing medicines. According to the American Medical Association (AMA), drug companies spend $4.5 billion on those advertisements.[205] There is a concern in the medical community that ads drive the demand for expensive treatments rather than less costly alternatives. One study showed that alternative drugs for the treatment of schizophrenia cost as much as $600 less per month compared to the brand name Zyprexa by Eli Lilly or Seroquel by AstraZeneca.[206]

In November 2015, the AMA supported a ban on direct-to-consumer advertising on prescription drugs, reflecting the physicians' concerns about the negative impact of commercially driven promotion and the role that marketing plays in fueling escalating drug prices.[207] One of the common complaints is that DTCPA is used to promote expensive drugs that may not offer significant benefits compared to alternative cheaper drugs. Not only will banning the DTCPA help reduce the proliferation of expensive drugs, but it will also save $4.5 billion on advertising costs that can help lower the cost of drugs.

The legacy healthcare delivery model as well as the Affordable Care Act (ACA), also known as Obamacare, primarily depends on the health insurance industry to solve the United States healthcare problem, even though the real issue is the rising cost of healthcare not the health insurance coverage.

Ten

HEALTH INSURANCE

The emerging health insurance model is slowly moving toward the basic premise of insurance, which is to provide compensation for unforeseen conditions in return for a monthly premium. Health insurance, as the name implies, will finally evolve into a traditional insurance product as it applies to other aspects of personal and business affairs, such as life, car, home, or business insurance. As this model grows, consumers will be empowered to control their health maintenance costs while health insurance covers major expenses such as surgeries or emergencies. Insurance is designed to pay for accidents, disasters, and catastrophes and other unforeseen conditions. The current health insurance model pays for the maintenance of health, taking the responsibility away from the individuals. The net result appears to be unchecked healthcare costs and higher insurance premiums.

Women make about 80 percent of the family's healthcare decisions.[208] However, 78 percent of the women do not trust their insurance providers.[209] This is the reason why women's leadership is so important for health insurance companies. These women are executives at the top ten health insurance companies in 2015: Karen S. Lynch, President of Aetna; Terry Bayer, COO of Molina Healthcare; Gloria McCarthy, Chief Administrative Officer at Anthem; Juanell Hefner, Chief Administrative Officer at HealthNet; and

Karen Ignagni, of America's Health Insurance Plans, a national trade association representing the health insurance companies that cover about 200 million Americans.

Employer-provided healthcare originated as a perk to attract workers and circumvent wage controls imposed during World War II. Eventually, it became a government mandate, but the result has been higher healthcare costs as providers received payments for treatment instead of preventions or successful outcomes.

Medicare and private insurers pay a highly-discounted price for medical procedures, and the difference between that and what hospitals and doctors charge (list price) results in high costs for patients who do not have any health insurance coverage. An approach like Smart Patient Credits (SPC) can bring free-market forces to bear greater efficiency. It will reverse the untenable rise in costs that threatens to put affordable healthcare hopelessly out of reach for an average American, with or without the Affordable Care Act.

Cost Control

Insurance companies spend 29 percent of their billing staff's time processing and appealing denials that are eventually paid, based on the Massachusetts General Physicians Organization (MGPO) study.[210] The power of claim denial or delay gives them the leverage to stay profitable. However, they can spend less on overhead by giving their subscribers incentives through SPCs as well as using big data to identify abuse or overuse of treatments.

Big data is available in the public domain through search engines, social media, and other venues that can identify patterns, trends, and any abnormalities from the routine medical condition. The routine condition claims can then be automatically approved by the insurance companies in real-time using big data to eliminate most of the overhead costs. For example, insurance companies can develop certain parameters for each individual member or even groups based on age, gender, health condition, and other related factors. Any claim that falls within those parameters can be approved automatically, saving money and time, avoiding potential medical complications and the patient's suffering. Of course, there would be exceptions to the rule and that is where decisions can be elevated to the professional staff.

The big data can also serve as a predictive platform where insurance companies can proactively inform their members to take precautions, such as asking members to get flu shots due to an increase in flu cases in their area. Through keyword search activities, Google was able to tell where the flu had spread and then was able to offer a timely indicator during the H1N1 crisis in 2009.[211] All of this data can be assembled and analyzed to make claim processing efficient and cost effective.

The Federal Government's decision to publish Medicare data on hospital charges in 2015[212] was long overdue. Without more widespread price transparency, consumers can't play their crucial free-market role in choosing reasonable prices or in assessing quality and value. Providing consumers with basic price information is just common-sense capitalism. However, after decades of chronic, institutionalized ignorance, consumers will need more than just clearly marked price tags to take up the kind of active role they typically play, and should play, when shopping for all kinds of goods or services.

Smart Patient Credits

Smart Patient Credits (SPCs) do not require any legislation or a government mandate, just a smarter application of free-market principles. It would work something like a credit card rewards program—except that the rewards would accrue from spending less rather than more.

SPCs would work this way: Insurance companies will provide to their subscribers the maximum dollar amounts they have agreed to pay to their healthcare providers for standard procedures and treatments. Moreover, providers would publish their list prices. If patients can get the service they want for less than the contracted price, they would receive SPCs for every dollar of savings—50 cents on the dollar, for example. So if a patient chooses a provider who charges $1,000 less than the maximum amount his or her insurer will pay, then he or she would receive $500 worth of SPCs from his or her insurer or Medicare. This could be applied to a copayment or another uncovered medical expense.

The California Public Employees' Retirement System (CalPERS) decision to cap the payment on certain standard procedures underscores the benefits of empowering the consumers. When Anthem Blue Cross members were told the company would cap its payments for a knee or hip replacement surgery at $30,000, some patients shopped around for hospitals that fit that budget. About forty higher-priced hospitals in California reduced their surgery prices to avoid losing patients while some patients simply chose less expensive hospitals. The average charge among the more expensive hospitals fell 37 percent from $43,308 in 2010 to $27,149 in 2012 for these common joint replacements. It resulted in savings of $5.5 million to CalPERS over a two-year period.[213]

This structure would give consumers a needed incentive to shop, evaluate, and decide which medical provider is best for them. By employing such incentives as SPCs, private insurers and Medicare would benefit by having their policyholders serve, in effect, as their auditors and cost controllers.

Health Insurance Reframed

The private insurance companies' goal is to maximize their profits. They work for their shareholders, not the patients. It would benefit the economy, the middle class, and Americans' health if the emerging health insurance model emulates the traditional insurance and warranty model. People can have catastrophic health insurance to deal with unforeseen medical conditions and buy a warranty (just like on cars or home appliances) for a period of time to take care of wear and tear of the body. Under this model, a patient would pay a small fixed fee to a healthcare provider for the visits, and the rest is covered through the warranty offered by a private company or a health insurance provider.

The emerging model will place personal responsibility at the front and center of the discussion. Every citizen should enjoy access to affordable healthcare. However, that is different than having 100 percent health insurance coverage for any health issue—especially for the unhealthy choices that many individuals make such as smoking, substance abuse, or poor eating habits. Under the current system, good personal habits are penalized since the health insurance premiums do not differentiate the healthy and unhealthy behavior. The result is that costs shift from high-risk patients to those who are low-risk, giving no incentive for personal responsibility.

The Affordable Care Act

The most popular and socially responsible components of the Affordable Care Act (ACA) are coverage of preexisting conditions, the spending cap, and coverage of kids until age 26 under their parents' policy. ACA is also known as Obamacare. Women seem to be the key beneficiaries: 7.7 million signed for health insurance coverage during the first enrollment. Moreover, 65 million women with preexisting conditions can no longer be denied the coverage, and 1.1 million women aged nineteen to twenty-five have coverage under their parent's plan.[214]

The fundamental issue still remains that millions of Americans are projected to be without health insurance even ten years after the implementation of ACA. Americans were led to believe that the Affordable Care Act (ACA) would end all of America's healthcare problems through health insurance coverage. It also increases the tax (called "penalty") if one does not buy health insurance. The cost of healthcare will continue to rise since the ACA does not address the cost of healthcare, despite its name. Moreover, millennials shoulder the bulk of the cost by having to buy health insurance they may not need in order to support the cost of those with medical conditions.

Former President Bill Clinton explained it this way, "This only works, for example, if young people show up (referring to Obamacare). …We got them in the pools, because otherwise all these projected low costs cannot be held if older people with preexisting conditions are disproportionately represented in any given state."[215]

Just like every well-intentioned federal program, the ACA has unintended consequences that undermine proposed benefits. According to the Congressional Budget Office (CBO) forecast, 20 million Americans are projected to be enrolled by the end of 2016, however the Department of *Health and Human Services in October 2015 announced that about* 10 million will be covered by late 2016, about half of what was projected earlier.[216] Furthermore, the ACA did not make any provisions to add more physicians or healthcare providers to accommodate these additional enrollees.

As a result of the ACA, some corporations are encouraging their employees to move to health exchanges instead of company health insurance providers to reduce costs. Walmart is hiring more part-timers to avoid the ACA mandate.[217] Other large corporations such as UPS and Delta Airlines are pulling back on healthcare benefits, citing higher costs due to the ACA, according to a CNNMoney report.[218]

The Consumer Operated and Oriented Plan (CO-OP) was created by the ACA as a nonprofit alternative to add competition and choices. However, twelve out of twenty-three co-ops have failed as of March 2016. These co-ops benefited from risk corridors—federal funds intended to subsidize them in case of sicker customers and bigger claims than anticipated—but the government only paid 12.5 percent of nearly $3 billion owed.[219] Competition may be further reduced as some of the largest health insurance companies such as Aetna and UnitedHealth consider no longer selling ACA coverage if they continue to lose money due to higher than expected claims for individual policies.[220]

The proposed benefits of the ACA could have been accomplished without starting another entitlement program that primarily depends on health insurance as the solution. Since the ACA is now the law of the land, all of these problems can be addressed through transformation instead of trying to repeal the law, as Republican lawmakers have been doing.

One way to make the ACA more efficient under the overall healthcare transformation umbrella is to consolidate all exchanges into one single entity as a public option that will compete with private health insurance companies for subscribers. That public option could be set up under Medicare: one group for retirees, as it currently exists, and a second new group for catastrophic insurance and warranty coverage for paid subscribers. By combining the purchasing power of both groups, a transformed Medicare could offer better coverage at a lower cost.

Eleven

MEDICARE

Medicare is one of the most successful government programs because it kept elderly people at home and out of poverty. Unfortunately, it will be insolvent by 2030 based on a 2015 report by the Social Security Administration.[221] That reality could serve as the impetus for the transformation of Medicare and the healthcare system at large. The transformation has started with the announcement by the Health and Human Services (HHS) Secretary Sylvia M. Burwell in January 2015 to move 90 percent of the current Medicare program to an alternative payment model that is based on results by 2018. Currently, alternative models represent only 20 percent of Medicare payments. This model did not even exist as of 2011, but it has shown significant results between 2010 and 2013 saving 50,000 lives and $12 billion in healthcare spending, according to the preliminary estimates by HHS.[222]

Medicare is a health insurance program for people age 65 or older as well as people with disabilities or permanent kidney failure. It has two components: Part A, which pays for hospitals, skilled nursing facilities, and hospice care; and Supplementary Medical Insurance (SMI), which consists of Medicare Part B and Part D. Part B helps pay for physicians, drugs, outpatient hospitals, home health, and other services, while Part D provides subsidized access to prescription drugs not covered under Part B.

The Medicare Act in 1965 provided economic security to women during old age where they had the greatest need for medical care. Women's influence is reflected in the expansion of the benefits provided under Medicare over time. In the early days, the benefit package was focused on hospital coverage. Preventive care and prescription drug benefits were not typically included as they are now. Fifty-six percent of those enrolled in Medicare are women,[223] so women will have a significant role in reshaping Medicare for the coming decades, as they have under the leadership of HHS Secretary Burwell.

Another encouraging sign is the transparency of Medicare in releasing information on payments to healthcare providers. The data summarizes the utilization and payments for procedures, services, and prescription drugs provided to Medicare beneficiaries by specific inpatient and outpatient hospitals, physicians, and other suppliers. This data includes information for the 100 most common inpatient services, thirty common outpatient services, all physician and other supplier procedures and services, and all Part D prescriptions. The data disclosure highlighted the abnormalities in payment as well as the weakness of the current model—all of which helps get consumers involved.

Medicare benefits payments totaled about $600 billion, or almost 20 percent of America's healthcare expenditure in 2013.[224] Being such a large part of national healthcare expenditure, Medicare must play a central part in curtailing the cost of healthcare in conjunction with the private sector. Employers are increasingly taking steps to move to more efficient healthcare delivery models. Medicare is transitioning away from the fee-for-service model toward alternative payment models that rewards improved outcomes and reduced costs.

The pay-for-performance model is the new alternative payment model in the healthcare sector. The provider of service gets paid according to performance. Broadly speaking, pay-for-performance incorporates any program or initiative that improves the quality and value of the healthcare and makes it more efficient. This is a major shift away from the current fee-for-service model. A fee-for-service model rewards healthcare providers for procedures and services regardless of the quality of service or benefit to the patients.

Under the current model healthcare providers receive payment for each service such as a visit to the doctor's office, surgery, or blood test, whether that service helped or hurt the patient. That results in more procedures and services being performed because that benefits the provider while increasing the cost.

HHS set a goal of tying 30 percent of the fee-for-service payments through Accountable Care Organizations (ACO) or bundled payment arrangements by the end of 2016. This is the first time in the history of Medicare that HHS has set explicit goals for alternative payment models and value-based payments. In this model, healthcare providers are accountable for the quality and cost of care delivered. They have incentives to reduce duplication of tests and screenings. This new initiative calls for an integrated healthcare model where ACOs could be groups of healthcare providers such as doctors, hospitals, and other providers who work together to provide a coordinated high quality healthcare to reduce the recurrence of the medical condition or readmission to the hospital.

Unfortunately, this model falls short of one key component: personal responsibility. Healthcare providers are responsible for the quality and the performance of the service. However, they have no control over the behavior of the patients who choose not to follow the drug protocol or behavioral changes required for a better outcome after their visit to a doctor or a hospital. That is where an incentive for patients, such as Smart Patient Credits (SPC) comes into play.

The HHS administration's attempt to transform Medicare from the fee-for-service model is a tacit acknowledgement of the failure of the current payment model. It is one of the causes of higher healthcare costs and involves fraud, waste, and restriction on competition. However, some simple solutions could make a big difference.

Alternative Drugs & Compounded Medications

Having a large bureaucracy like Medicare with Congress micromanaging it adds cost by making the system inflexible. For example, the drugs Avastin and Lucentis, according to the research, are both effective and have fairly comparable side effects when used to treat the wet form of macular degeneration.[225] Avastin costs $55 per treatment and Lucentis costs $2,023 per dose, but the

FDA approves Lucentis for the condition and Avastin is approved for cancer treatment only. Even though both drugs are approved by the FDA, they cannot be substituted from one condition to another. However, a large number of doctors use cheaper Avastin for the ocular condition as the off-label drug—one prescribed for a use other than the originally intended one and repackaged by compounding pharmacies. This swap could save Medicare $18 billion and $4.6 billion in copay to patients.

Medicare generally does not pay for compounded drugs but some insurance companies do cover them. The forty-to-one price discrepancy between Avastin and Lucentis can be attributed to the repacking of the drug by compounding pharmacies. They take a 4 ml single dose vial of Avastin and divide that into as many as sixty small dosages—enough to treat the ocular condition depending on the techniques used by the ophthalmologists. On the other hand, Lucentis is supplied in the volume needed for a single use at a much higher cost.[226] By allowing compounding medicines to be used as a replacement for expensive branded drugs, Medicare can incur significant savings. On the other hand, generic drugs saved $76 billion to Medicare and as much as $1,923 per enrollee in 2014.[227]

Price Negotiation

In 2003, Congress prohibited Medicare from negotiating the drug prices. President Obama, during his 2008 campaign, decried the cozy relationship between Washington and the drug lobby. In one of his political ads he said, "The pharmaceutical industry wrote into the prescription drug plan that Medicare could not negotiate with drug companies. And you know what, the chairman of the committee, who pushed the law through, went to work for the pharmaceutical industry making $2 million a year. I don't want to learn how to play the game better; I want to end the game plan."[228] Unfortunately, despite the ACA, Medicare still cannot negotiate the drug prices.

In 2013, Medicare Part D alone spent nearly $70 billion in prescription drugs. Even with such a large purchasing power, Medicare paid 198 percent of the median costs of the same brand-name drugs in the thirty-one countries that are part of the Organization for Economic Co-operation and Development

(OECD). Furthermore, Medicare gets only 16 percent discount on average from the official manufacturer price while the Veteran Health Administration (VHA) gets a 54 percent discount and Medicaid a 52 percent discount both with significantly lower drug budgets, according to a policy brief by Carleton University.[229] A brand name drug priced at $100 will cost Medicare $83, but it only costs Medicaid $48 and the VHA $46. The nation could have realized a savings of $16 billion in 2010 alone if Medicare had the authority to negotiate the same price as the VHA. In the private sector, companies like Costco or Walmart could've squeezed a much higher discount with Medicare's purchasing power of $70 billion.

Fraud

In 2013, Medicare served 51 million Americans with a budget of $600 billion and a payment model that had no accountability for results or performance. One of the outcomes of such an inefficient model was $50 billion in fraudulent payments, according to a Government Accountability Office testimony to Congress in April 2014.[230] Medicare's published data in 2012 underscored fraud concerns: a Florida ophthalmologist took in more than $26 million to treat fewer than 900 patients, and a cardiologist received $23 million—nearly 80 times the average payment for that specialty.[231]

One of the best ways to curtail fraud is the use of big data, as mentioned in the health insurance chapter, to detect abnormalities before any reimbursements are made. Furthermore, having the patients attesting the bills to confirm the services rendered and the charges submitted would also hinder the submission of fraudulent charges. Another way to reduce fraud would be to empower the Medicare patients and provide them with incentives such as Smart Patient Credits.

Twelve

United States Education System

Technology and the sky-high tuition costs offer a fertile ground for disrupters. If the American economy can bring secondary and higher education in sync with the needs of employers, over five million unfilled jobs could easily be filled. Moreover, employers would end up clamoring to hire qualified workers, offer flexibility and benefits as enticements.

Women are already prominent in K-12 education. Teachers serve as role models for students and help shape students' characters and values during their formative years—76 percent of K-12 teachers[232] and 52 percent principals are women.[233] Four of the eight Ivy League universities have women presidents, including Drew Faust at Harvard, Amy Gutmann at University of Pennsylvania, Christina Paxson at Brown University, and Elizabeth Garrett at Cornell. Women head ten out of twenty-three California State Universities with a total of 460,000 students.[234] Nationwide only 26 percent of college presidents[235] are women, which is an opportunity for the 48 percent of college faculty who are women to gain further ground.

The position of Chief Academic Officer (CAO) serves as the stepping-stone for the presidency of a college. Women hold about 40 percent of the

CAO positions at four-year colleges, which bodes well for the women's ascendency toward university presidency and influence on the future of education in America.[236]

America is the envy of the world for its network of colleges and universities, a global magnet for higher education. People from around the world want to send their children to universities and colleges in the United States. Nearly 400 heads of foreign governments and seventy-seven Noble Prize winners have been trained or educated in the United States as of 2015, according the United States State Department.[237]

It is in America's strategic interest to transform the educational system to meet society's long-term needs and be competitive in the twenty-first century. The nation needs a flexible, student-centric, interactive system that serves the students of tomorrow—65 percent of today's grade school students will end up in jobs that have not even been invented yet.[238/239]

America's one-hundred-year-old educational model is going through a transformation thanks to the Internet and innovative disrupters. The old model worked well in the last century but it is not meeting the needs of the future, as reflected by the millions of unfilled job openings. One of the challenges is to redefine the purpose of education.

Education Redefined

Experts, parents, and students are reflecting upon the purpose of education in light of ever increasing cost. Why do students need four years of college to earn a bachelor's degree? Why can't the same proficiency be accomplished in two or three years (at least in certain fields)? Why do universities emphasize research, while undergraduate students—the bulk of higher education pupils—go there to prepare for occupations? What happens to those whose aptitude is to work with their hands and machines and would prefer practical experience rather than theories? Why do doctors need to spend eight years on higher education after high school instead of five before they can practice their profession? These questions are instigating a paradigm shift in education.

Historically, the system's goal has been to teach students' curriculum in an efficient way,[240] resulting in a *production model*. The United States presently spends significant amounts of money on an assembly line education system, using the standardization and production models of passive students who listen to teachers' lectures all day long, five days a week. Unfortunately, very little if any time has been spent to ensure that students actually learn or acquire skills—and to discover how life-changing that process can be. Curricula can be standardized, but *learning* cannot be standardized, since no two brains are the same, contends Salman Khan of the Khan Academy.[241] Traditionally, society has viewed a college education as a gateway to employment,[242] but 45 percent of 2012 college graduates between the age of twenty-two and twenty-seven were unemployed or underemployed as of 2012.[243]

Is the goal of education to build an educational system or to have students actually learn something?[244] To meet the demands of the current and future job market and economy, the twenty-first century education system needs to teach students to learn and acquire skills, but more importantly to inspire and teach lifelong learning skills. Learning to teach new skills will keep workers productive and relevant in an ever-changing technological age and global economy.

Cost of Education

Improved efficiency for K-12 education begins with making sure that teachers are highly compensated. Studies have shown that the countries that rank high in education tend to offer teachers higher status in society. In society, like it or not, status and pay go together.

The challenges go back to two well-intentioned federal goals: first, a college education should be within the reach of every American, and second, that if students borrow money from the federal government, they should repay it. Most of us would agree that both are noble goals but the consequences of both have been stunning. According to a Federal Reserve Bank of New York's staff report there is a pass-through effect on tuition of Pell Grants and subsidized direct loans of about 55 and 65 cents on the dollar respectively.[245] In other words, colleges raise tuition by 55 to 65 cents for every dollar available to students through direct or indirect government loans.

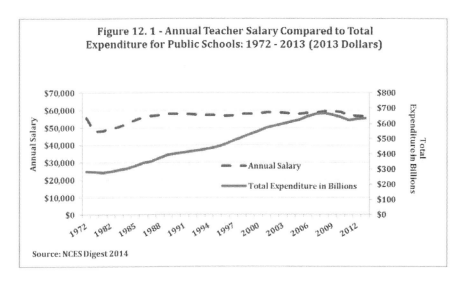

Figure 12. 1 - Annual Teacher Salary Compared to Total Expenditure for Public Schools: 1972 - 2013 (2013 Dollars)

Source: NCES Digest 2014

America continues to spend more and more money on K-12 and higher education but seems to be getting less and less in return. The real issue is misallocation of resources and inefficiencies that have been built over a period throughout the system. Teachers' salaries have remained about the same since 1972, but the total expenditure has more than doubled over the same period (Fig.12.1).[246] Enrollment in public schools over the last 15 years has remained about the same, at about fifty million, and is projected to have very little growth for the next ten years.[247] However, expenditures on education do not appear to be adjusting to the slowing growth of enrollment.

Another factor hindering quality education is that the Federal Government exerts control through mandates or incentives, such as No Child Left Behind or Race to the Top, even though it only contributes just

9 percent toward education. Most of the funding comes from local and state governments (Fig. 12.2).[248]

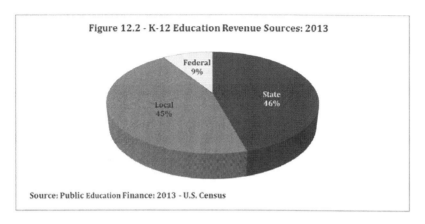

Figure 12.2 - K-12 Education Revenue Sources: 2013

Federal 9%

State 46%

Local 45%

Source: Public Education Finance: 2013 - U.S. Census

Across the nation, people and governments are trying to find ways to maximize the capacity and performance of the education system. As a result, the public's attention is moving toward teacher quality, accountability, and the pay-for-performance model. Pay-for-performance means high performing teachers should be better compensated than poor performing ones.

To encourage that paradigm shift, the United States education could be paying teachers' salaries that show a greater regard for the work they do. Higher compensation will also attract talent that might otherwise seek different professions. With higher pay goes responsibility and accountability. The emerging model will move toward decentralizing decision making power so administrators can terminate teachers who fail to perform and are unable to serve the needs of the students. This would alleviate the current issue of sending poor-performing teachers to already poor-performing schools, such as those in the inner-city[249] to avoid a long drawn out firing process as is often the case under the current system.

At the same time, higher education is quickly growing out of reach for lower and middle-class Americans. Transformation of the higher education delivery model is moving forward as this book is written. It is starting by simply focusing on some basics, such as why students are required to physically attend a college class, especially general education courses that have 400 students in a classroom. More and more of those subjects requiring little student–teacher interaction are being taught online. By encouraging long distance learning through online classes, universities can free up expensive real estate and resources, improve productivity, and increase the size of their student population significantly. This will make college costs accessible to tens of thousands of additional students.

Student-Centric Education

Interactive learning is more exciting and fun than passive learning. Today's K-12 students have grown up with video games, mobile devices, and the Internet, so they are less adept at listening passively for a long duration. From a very early age, children are accustomed to interaction.

Furthermore, the education system is recognizing more and more that human beings are complex, and each person is unique in his or her understanding of information and ways of learning. Some people are visual while others are auditory or kinesthetic, learning through a hands-on approach. Some have aptitudes for science and technology, while others enjoy tinkering with their hands and want to be mechanics or build things. Technology is making personalized learning possible that did not exist twenty years ago.

Another major paradigm shift is the realization that not everyone benefits from college education, yet anyone can be successful and contribute to society, whether as construction workers, car mechanics, or security guards. Germany focuses on matching those students who are not inclined to go through the typical college education with the vocation that best fits them. Their education system has partnerships with employers, unions, and the government for matching and providing the necessary vocational training. The United States lacks a mechanism to match individuals with the right vocation and help get them the training through vocational schools or apprenticeships and internship with potential employers. That mechanism is evolving through technology and social entrepreneurs.

Technology is increasing efficiency and productivity, but it can also be disruptive unless there is an environment of ongoing learning. Yesterday's jobs are being replaced by technology, as has been observed over the last few decades, whether it be customer service, airline check-ins, self-service gas stations, self-service checkouts at grocery and retail stores, and so on. Therefore, higher education needs to be focused on equipping students for change through self-learning and permanent learning models.

There is a sense of urgency and acknowledgement that the current situation requires major overhaul and that a flexible system that serves future students is a must. The new approach should be able to meet the challenges of technological evolution. In this model, teachers serve as coaches who guide the students to learn on their own. They are facilitators rather than pedagogues. Learning should be the responsibility of the students and not just of the teachers. Students will be taught how to teach themselves. The reality is that no one can predict what people will need to know in 20 years in order to be productive members of society. Therefore, the best way to equip students is by teaching them how to learn. This will enable them to effectively adapt for the future.

Thirteen

K-12 Education

Innovative disrupters are transforming education for elementary, middle, and high school students. The emerging reality allows for higher teacher compensation through large classes that harness the power of student-centered learning.

The current K-12 education model is based on the eighteenth century Prussian model, featuring publicly funded compulsory education.[250] The Prussian model required a national curriculum for each grade, training for teachers, and compulsory attendance and testing for students. Specific subjects were taught during designated class periods. The model was designed to educate students on a large scale in an economical way. Later, the Prussian model came to America via Massachusetts through Horace Mann, then-Massachusetts Secretary of Education. He was also known in the middle of the nineteenth century as the "father of public education."[251]

This model has not adapted to the changes of society and the economy. Thus, its return on investment and educational effectiveness seems to be diminished. In 2011, America's annual expense of $13,210 per pupil had doubled from $6,644 in 1972 (Figure 13.1).[252] Yet, according to a Harvard University report, high school graduates scored at a 32 percent proficiency rate in mathematics. All of those countries whose students score better spend less per student than the United States does, except for Switzerland. Improving these results would benefit the American middle class: if America's students were to improve their math and reading proficiency over South Korea's ranking, the national income would rise by an estimated $1 trillion per year, according to the same study.[253] Forty-seven percent of South Koreans were proficient in reading compared to 31 percent of Americans while in math they surpassed the United States 57 percent to 32 percent.

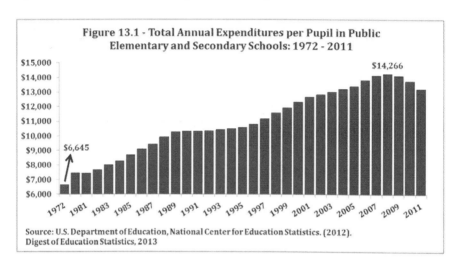

Figure 13.1 - Total Annual Expenditures per Pupil in Public Elementary and Secondary Schools: 1972 - 2011

Source: U.S. Department of Education, National Center for Education Statistics. (2012). Digest of Education Statistics, 2013

Moving Forward in K-12 Education

Women play a critical role in K-12 education; in 2012, 52 percent of principals were women[254] and 76 percent of teachers were women. For example, Michelle King was selected as the first African-American woman district superintendent by the Los Angeles Unified School District board in January 2016. She spent 31 years as a teacher and principal with LAUSD.[255]

New paradigms and technological ideas are taking shape to prepare kids for the global economy from an early age. Corporations and social entrepreneurs are taking the lead, rather than relying on the government, as they recognize the need for a more appropriately educated labor force. The Gates Foundation, Khan Academy, Oracle Corporation, and others are some of the examples. Future learning will be self-directed. Rather than listening to lectures, students will progress at their own pace while older students will mentor the younger ones. Teachers will co-teach a larger class of 100 or more students from different grade levels. Their role will be to act as coaches and facilitators where they will pay one-on-one attention to those students who are falling behind.

Philanthropists have poured millions of dollars into the United States public schools attempting to reform it without much success. Rigid bureaucracies, centralized control, and politics seem to be the biggest hurdles these reformation efforts face. For example, in 2010, Mark Zuckerberg donated $100 million to Newark, New Jersey's failing public school system in order for the system to serve as a model for education reform over five years. That $100 million was supposed to be matched by other sources to make it a $200 million proposition. More than half of that money went toward labor and contract costs ($89 million) and consultants ($21 million), some of whom were paid $1,000 a day, instead of that money being spent on students. Zuckerberg wanted the flexibility in rewarding high performing teachers and firing non-performing ones in teacher contracts and allocated $50 million of his donation toward that goal. However, since the New Jersey Legislature determines those types of protection, he was unable to achieve that goal.[256]

The Los Angeles Unified School District (LAUSD) had a plan where every public school student in Los Angeles would have an iPad to replace hardcopy textbooks. The program, billed to cost $1.3 billion, was abandoned two years later in 2015 because it was too expensive and unsustainable. LAUSD Superintendent Ramon Cortines characterized the district as lacking an instructional plan to use the technology.[257]

Some major foundations in Los Angeles, led by the Broad Foundation, are working around the bureaucracy by placing their money and efforts on charter schools in Los Angeles. There is a discussion of investing $490 million with a goal of enrolling half of the Los Angeles Unified School District's students over the next eight years. It is a bold plan, however, and it will face strong opposition from many city labor unions which may render it ineffective.[258] Regardless, all these efforts are challenging the status quo, a good first start toward transformation.

Silicon Valley, the hub of disruptive innovations, is taking a different route realizing that innovations are inherently risky. One of the ways it is minimizing the risk is by keeping the schools private or offering alternative ways of learning. One cannot blame the parents or educators who do not want their children to be the guinea pigs. Therefore, they are building their own alternative solutions to the current centralized top-down model.

Altschool raised $100 million from Silicon Valley elites to create micro-schools that offer a personalized learning experience. It is an odd blend of retro and futuristic—Montessori 2.0, according to its founder Max Ventilla, a Google alumni.[259] Each school is a single mixed-age class of 25 to 30 kids with two teachers, hence the name micro-schools. One can spot the 12- and 13-year-olds doing yoga while younger students are working with tablets and laptops to complete personalized lessons and projects. The course content may include activities such as a 20 minute math lesson on Khan Academy's website.[260]

Khan Academy, a nonprofit educational organization, was founded in 2006 by Salman Khan. Khan Academy's goal is changing education for the better by providing a free, world-class education for anyone anywhere.[261] Their tag line is, "You can learn anything. For free. For everyone. Forever." Khan's vision is to move away from the daily schedule of 50-minute periods, grades, and teaching all students the same material at the same pace, because that old classroom model does not meet their changing needs. It's a fundamentally passive way of learning while the world requires more and more *active* processing of information.

Khan Academy offers more than 6,000 educational videos across different platforms and subjects—half of them are about math and science. Their model is to have teachers send online lectures for students to access at home and then use the class time for discussion, practice, and group projects. More than 100 students of various ages can work at their own pace being monitored by multiple teachers in the room. The teachers' role is to make sure that students are making progress and provide assistance rather than lecture.

The site had more than eight million unique visitors in September 2015[262] and has delivered more than 580 million video lessons on-demand. It boasts over one million teachers around the globe incorporating the supplemental teaching tool into their classrooms.[263] Microsoft founder Bill Gates used them for his kids. He donated millions to Khan Academy through the Gates Foundation.

The Gates Foundation, through its College Ready Strategy, is focused on the teacher-student bond. Their goal is to prepare *all* students for college and careers. The premise is that at the heart of learning is the bond between the teacher and student, regardless of the class size. The bond helps trigger a student's hunger to learn and allows teachers to guide each student's individual path. This initiative provides grants to partner organizations to develop Smarter School Spending, a free online source to fund their initiatives, and charter collaboration. Partner organizations could be big or small and for-profit or not-for-profit.[264]

Oracle, one of the top software companies in Northern California, is building a tuition-free, state-of-the-art high school on its corporate campus. According to its founder and CEO, Larry Ellison, it's a school where students "learn to think." It will be a charter school as part of the San Mateo High School District devoted to science and technology. The Design Tech High School, or d.tech for short, will open in 2017 and Oracle employees will volunteer their time by teaching workshops on subjects like coding and design and will work closely with students on projects related to STEM (science, technology, engineering, and mathematics).[265] The hallmark of the program is its personalized learning experience, active interdisciplinary learning, real-world engagement, and problem solving. Daily activities will involve traditional classes combined with personalized activities.

Teacherspayteachers.com (TpT) is another interesting platform that helps enhance teachers' productivity by allowing them to share, sell, and buy educational material from each other. Founded in 2006 by a New York City public school teacher, it is one of the first open marketplaces as a resource for teachers. The company estimates that, in August 2014, one in three United States teachers downloaded learning materials from their website. They have 1.7 million available resources with 3.5 million global members.[266]

Technological advances, like Google's virtual reality system, can offer field trips specifically designed for students based on course curriculum. An economics teacher at Saint Francis High School in Mountain View, California, used this system to create a "Great Recession Tour" of Manhattan, New York. This virtual reality system took the students to the headquarters and offices of Lehman Brothers, Goldman Sachs, and federal regulators involved in the 2007-08 financial crisis.[267] In addition, Google's Apps for Education offers email, a calendar, and document-sharing products for free to schools and is being used by 45 million students and teachers.

The real-world focus of these programs allows for partnerships with corporations. They are one of the key stakeholders in the education system because they're the prospective employers of graduates. One successful example is Brooklyn P-TECH, created by IBM in partnership with the New York City Department of Education and the City University of New York. Their curriculum is designed to meet the industry needs, provide mentors and paid internships. They place successful graduates first in line for jobs at IBM and other participating companies. There are sixty such programs across six states that will be open by 2017.[268]

These projects and ideas are relatively small in the scheme of things, but they are positive signs toward a more efficient and less expensive (or even free) education for all. The pace of these transformational ideas will grow as technology develops and more disruptive innovators dedicate their talents to the cause.

Fourteen

Higher Education

As the higher education system transforms to meet the challenges of the twenty-first century, there will likely be much lower tuition costs, and students will spend more time getting real-life experience outside the classroom. Moreover, economic demand will dictate employers to share the cost of educating their employees.

There are three broad categories of universities. Some are major research institutions with doctoral programs. They are arguably known more for their research than their teaching. Professors get tenured not necessarily because they are good instructors but because of their research and publications as well as for their years of service. Moreover, some of these universities view themselves first and foremost as a place for intellectual and social experiences; employment for graduates is a secondary consideration. Some are focused on undergraduate studies like the California State University system. The third category is a mix of research and preparation for employments like Loyola Marymount in Los Angeles, a private liberal arts college. Students for a four-year college generally select their place of higher education based on employment prospects first and consider the intellectual experience second.[269]

Just like in the K-12 model, universities suffer from misallocated incentives and investments: salaries and wages for instruction were only around 26 to 28 percent of total expenditures at public degree-granting institutions from 2005 to 2011(Fig. 14.1).[270]

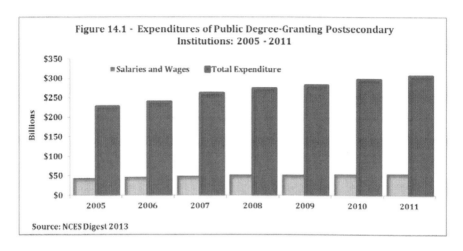

Figure 14.1 - Expenditures of Public Degree-Granting Postsecondary Institutions: 2005 - 2011

Source: NCES Digest 2013

Women have made great strides in college education over the last 30 years. From 1970 to 2009, women earning bachelor's degrees grew from 43 percent to 57 percent, master's degrees from 40 percent to 60 percent, and doctorate degrees from 13 percent to 52 percent. Women are getting closer to parity with men as full-time faculty at degree-granting institutions, reaching 43 percent. Moreover, women outperform men 56 percent to 44 percent in national research awards and grants.[271]

A typical college president is white, male, Protestant, married with children and holds a doctorate in education.[272] The number of college presidents sixty years and older increased to 60 percent in 2012, making way for a wave of retirements.[273] Women head many high-ranking universities: Linda Katehi at University of California Davis, Ana Mari Cauce at University of Washington, Phyllis Wise at University of Illinois at Urbana-Champaign, Rebecca Blank at University of Wisconsin at Madison—all of these universities were ranked in the top ten by Times Higher Education of U.K.[274]

Reinventing the College Model

Various models from boot camps, online learning, hybrid models and vocational training for some are the emerging new educational training models. They will continue to evolve as college tuition increases and as technology develops. Some thinkers are even challenging the notion of the current degree structure by advocating for a more flexible model for a new breed of students.[275] The result will be a system that better prepares students to evolve with the needs of the market and ensures that employers can find qualified workers to meet the demands of their business.

College for America (CFA) is partnering with employers such as GAP, McDonalds, and BlueCross to offer a fully accredited degree program to employees for $2,500 a year all-inclusive tuition. CFA is part of Southern New Hampshire University. Their model is designed for working professionals and is a project based learning model instead of traditional lectures and exams. Students work with a flexible online schedule at their convenience and pace.

Coursera, a Silicon Valley based for-profit corporation, offers an education platform that partners with over 120 top universities and organizations worldwide to provide courses for anyone. It is one of the leading providers of massive open online courses (MOOCs) along with other companies such as Udacity and edX. The courses are free for anyone, except certificates cost a modest fee of $30 to $100 per course.

Udacity, born out of Stanford University, offers online courses for free. Its mission is to bring accessible, affordable, engaging and effective higher education to the world. Udacity, in partnerships with Silicon Valley tech giants like Google, Facebook, and others, offer Nanodegree programs. It offers credentials in web development, data analysis, and programming through online courses and hands-on projects. The program takes about six to twelve months to complete the requirements at a cost of $200 per month. However, students get half of their tuition back when they graduate. Moreover, courses are free for everyone to use unless one wants a Nanodegree credential. edX was founded at Harvard and MIT in 2012 as an online learning destination and a MOOC provider, offering high quality courses from the world's best universities and institutions to anybody who wants to learn. It is an open-source platform for all educators to build learning tools for students. It also offers free online courses on a whole range of subjects. However, there is a small fee of less than $100 for a verified certificate of achievement after the completion of the course. They have partnered with Arizona State University, where students can earn credits for certain edX courses toward a traditional degree program.

Numerous colleges and universities are offering online degree programs at a significantly lower cost. Coursera has partnered with the University of Illinois to offer an MBA program. It is designed for working professionals for a total cost of around $20,000, which is significantly lower than the equivalent traditional college graduate degree programs. The Massachusetts Institute of Technology (MIT) offers free online courses. It also offers a MicroMaster's credential for $1,500, and a full degree program in supply chain management at half the cost and time compared to a traditional yearlong program.[276] Arizona State, Georgia Tech and others are following suit as these types of programs become more prominent.

Some colleges are reducing their fees to make higher education available to more families and students. More families ruled out colleges for their kids based on price—69 percent in 2012 compared to 53 percent in 2009. Some of the small colleges like Rosemont in Pennsylvania and Utica in New York are moving away from a high-price, high-discount model to a low-cost, low-discount model, reducing the list price of tuition by more than 40 percent. Normally, students do not pay the full tuition or list price once they take grants, loans and scholarships into account. At Rosemount, nearly 90 percent of freshmen pay a discounted price.[277] Some parents don't allow their children to consider prospective schools due to higher list prices. Therefore, some schools are providing discounted prices upfront to minimize the loss of potential students. Seton Hall University, the University of Chicago, and the University of Cincinnati, among others, have also lowered their tuition. The University of Chicago, one of the top ten universities in the country, has pledged to allow Chicago students to attend the university debt free, a savings of over $40,000 per year.[278]

Emerging models embrace a stronger value proposition. They balance the cost of education with a skills-based rather than credential-based education that students need to thrive in today's economy. The limitations of the present system show up in unfilled jobs, and low graduation rates: 59 percent of full-time undergraduate students who began school in 2007 graduated by 2013, or 41 percent of the students did not graduate within 6 years or at all.[279] The new models are more specialized or are implemented as a hybrid between traditional and online learning. Others are short-term with specialization in programming or are demand based in the form of skills-based learning.

Skills-Based Learning

Boot camps are short-term (four to twelve weeks) specialized programs designed to meet students' needs. Most of the boot camps are for-profit. They are mostly focused on programming and web sectors. However, there appears to be a growing interest in other sectors where traditional colleges do not offer degree programs. The advantage of this model is their adaptability to the market demand. They can create new curriculum and courses within weeks or months.

Teachable.com is another model for decentralization of education where anybody can offer a course on any subject at a nominal fee. It will not be too long before more well-known professors offer courses related to their particular expertise to students around the world at less than $5 per course. Imagine having 100,000 or more students from around the world taking a course from an industry expert.

While the new affordable higher learning models are being tested and perfected, the basic economic concept is also in play, where supply and demand for higher education under the current legacy model will bring the tuition cost down even further.

Integrated Education Model

An integrated higher education model, in conjunction with the legacy college system, can reduce college costs by half through efficient use of physical assets and technology. This model will combine the theoretical concepts and experiential learning as part of the graduating curriculum along with virtual learning whenever possible. In this model, for a four-year bachelor's program, students would only have to be physically present for two years or less. In addition, they will spend two years in an apprenticeship, study abroad, and internships in Washington or state capitals. This will double the physical capacity of colleges and allow them to nearly double enrollment. In return, colleges may be able to cut the cost by as much as half or more.

The two years of outside experience will better prepare students for their careers. Traditionally, apprenticeship is tied to a craft. However, it can expand to all fields since all students have to move to the real world after graduation. Basic wages earned through apprenticeships may pay for the study abroad program. The younger generation is financing Social Security, Medicare, and similar programs for the old. Therefore, it is important for them to be part of the decision-making process in Washington and the state capitals that affect their future. An internship in federal or state capital will get them civically engaged. Studying abroad will provide an appreciation of American lifestyles and culture. It will help students experience how other people live, think, and learn their perspective. This could be one of the best ways to bring people together, which is critical in a global economy. The real-world focus of these programs allow for partnerships with corporations, who are stakeholders in the education system.

Two transformational ideas are proposed by Salman Khan of the Khan Academy; creating a universal degree that is comparable to a Stanford degree and transforming the college transcript into a portfolio of artifacts that students have created, known as Universal Credentialing and Portfolio.[280]

Fewer and fewer students can afford to go to college under the current system, making it hard to get a degree. Khan is proposing a universal credentialing system that could compare a graduate of "Stanford or Harvard" by their raw abilities. The concept is to measure the abilities of all students and therefore make the granting institution irrelevant. To that end, Udacity and Coursera, both mentioned earlier, are moving in that direction with their Nanodegree and "signature track" certificates respectively.

The traditional job prospecting model is based on one's degree, college name recognition, and GPA. A degree reflects a certification of expertise or that a degree holder knows something about that subject or the field. However, that does not necessarily reflect the true skill sets required for today's job market or what a fresh graduate can do for the employer. That is why companies like Google do not care about college degrees.[281] That leads to the concept of a student's portfolio of products to show his or her ability to learn and adapt to any new reality in the marketplace.

While the United States' higher education system is going through a transformation, its legacy continues at least in one form through the accumulation of more than one trillion dollars in student loans.

Fifteen

Student Loans & Housing

Student loans are one of the main hurdles for many first-time homebuyers. More than $1 trillion in student loans, affecting forty million people,[282] has been a drag on economic growth, particularly the housing market. Historically, the answer to declining real wages has been a need for a college degree. However, the cost of higher education has skyrocketed to a point where fewer families can afford to pay, which has resulted in the ballooning of student loans. The good news is that the student loans for the next generation of college students may not be as big of a problem as the economy grows and the newer models for higher education emerge.

One factor behind the rise of student debt is the declining rate of labor participation by students while in school since the 1980s. In the eighties, students combined studies with part-time work to partially pay for college. However, that trend has been changing over the last two decades: fewer eighteen to twenty-four-year-olds enrolled in school are participating in the workforce.[283] The net result is the increasing amount of student loans to pay for the tuition.

Housing is the largest share of middle-class household assets.[284] A dream of homeownership—which became a nightmare for many during the housing crisis—still remains one of the main aspirations of the American middle class. However, with shifting demographics and economic realities, many middle-class Americans may not think of it as a high priority. After many Americans borrowed beyond their means, they are cautious about borrowing and saving more. The percentage of home sales that were distressed properties is down to 7 percent from its peak of 40 percent during the 2007 housing crisis. The implication is that there are 33 percent fewer distressed properties in the market to be sold as foreclosures.[285] In addition, household formation has increased to over a million during 2013-14, up from 580,000 during the 2010-13. This bodes well for the housing market since an increasing number of Americans are gaining financial stability to form a household.[286] A household formation is defined when an adult leaves the home of another adult to move into his or her own place, such as millennials leaving their parents' home.

Women account for 91 percent of home purchasing decisions.[287] Single women are buying homes at twice the rate of men[288] while the overall share of women homeowners is expected to grow as they surpass men in the workforce. Women take the same amount of loans as men do, but more of them pay a higher amount toward their debt payment—53 percent compared to 39 percent for men.[289]

Shifts in the economy and the makeup of the workforce are beginning to work in favor of the housing market. The first wave of millennial graduates, who have been in the workforce for a few years, are paying down their student loans. In addition, employers who are trying to find skilled workers to fill in millions of jobs in the United States are stepping up to the plate and offering to pay toward student loans. Some of the states, like New York, are offering a student loan forgiveness program; its "Get on Your Feet" loan forgiveness program will make up to two years' worth of loan payments for its residents who meet certain requirements.[290]

In 1980, Americans married at age twenty-two on average and had their first child by age twenty-three. That average has moved up to twenty-seven years of age by 2015 for marriage.[291] As the oldest millennials form families, they are more likely to purchase their first home. Higher rent in major cities could accelerate the pace of their purchase of homes in the areas where monthly mortgage payments after tax would be the same as the rent.

From lower interest rates to employer assistance programs there are a number of initiatives to help the middle class pay student loans.

Interest Rates

One of the easiest and the simplest solutions to the problem middle-class household's face, although it may be a politically difficult one, is lowering interest rates on student loans. The student loans rate offered by the federal government between 2008 and 2014 has varied from 3.4 to 8.5 percent depending on the loan programs, and can go as high as 10.5 percent.

The federal government owns about $1 trillion of the student debts[292] and makes a profit on these loans since it charges a higher rate while its cost is much lower—around 2 percent in 2015 measured by the ten-year treasury bills rate. The federal government booked almost $41.3 billion in student loan profits in 2013. To put it in perspective, these profits were the highest profits of any entity except for the two companies in the world, Exxon Mobile with $44.9 billion in 2012 and Apple with $41.7 billion.[293] Some will argue that this is an accounting matter and that the government does not make profits, but the facts prove otherwise. Lowering the student loan rates to match the cost to the federal government would put that $41.3 billion, which is probably more by 2017, into the students' pockets.

One of the shortcomings of student loans is having virtually no option of refinancing high interest student loans. This option has been commonly available for housing but not for student loans. New startups are looking to capitalize on this potentially $1 trillion market. Therefore, students can refinance their loans through private lenders such as SoFi and CommonBond, among a dozen others, and the competition will grow as traditional financial institutions see an opportunity for new customers and relationships.[294] The rates can start as low as 3.5 percent compared to some student loans that are in the 6 to 8 percent range thus reducing the interest cost by as much as 50 percent.

Financial Wellness Benefits

Employers want their workers to be as productive as possible. Many recognize that one of the factors that adversely impacts productivity is financial distress faced by their employees. In response, financial wellness benefits (FWB) are designed to enhance employees' overall financial well-being.[295]

About 80 percent of potential employees with student loans want to work for a company that offers repayment assistance—with matching opportunity. In response to the market demand for talents, companies who are desperate for new hires are offering assistance in student loan repayment as part of their FWB. They realize that it could save them as much as $3 for each dollar invested toward this benefit due to productivity gains and reduced medical costs and less stress.[296] By offering this benefit, employers are hoping to get a leg up in the hiring process and keeping the new talents loyal to them since this benefit only applies as long as they work for the company. The goal is to help employees so that they will have less worry and thus be more focused and productive on the job.

Some of the companies are contributing from $50 to $200 per month to help pay down the student loans. The consulting firm PWC announced in September 2015 that they will pay $1,200 per person per year to toward student debt. ChowNow, an online food ordering firm, offers $1,000 a year toward an employee's student loan payment.[297]

Income-Driven Payment Plan

The Federal Government offers income-based programs to reduce the student loan payment burden. One such program is the income-based repayment plan (IBR), which is capped at 15 percent of a graduate's discretionary income and any balance that exists after twenty-five years is forgiven. Discretionary income is defined as the difference between an individual's income and 150 percent of the poverty guideline for the family size and state of residence. There are similar programs with some variations, such as pay as you earn plans or income-contingent repayment (ICR).[298]

While these programs are helpful, the crisis graduates face could've been prevented. Well-intended student loan programs from the federal government made it easier for students to borrow even though they may not be able to pay after graduation. The idea was born due to the government's goal of making college education within reach for every American. Public Colleges and universities are incentivized to enroll as many students as possible so that they get the most federal loan dollars.[299] Therefore, colleges can raise tuition knowing that the money is easily and readily available even though their cost of educating a student has not gone up by that much.[300] Moreover, universities face no negative consequence when loans default since taxpayers take the risks, not the colleges.

Economic Impact of Student Loans

As the student loan problems resolve, the net result will be economic growth, which will impact housing construction, household formation, and consumer spending.

Alleviating student debt will allow more people to qualify for mortgages, resulting in higher demand for houses. Higher demand for housing will mean more need for heavy equipment, lumber, steel, electrical fixtures, cement and other industrial products that further the economic growth.

Household formation, another casualty of the student debt crisis, will improve as new graduates feel more financially secure with less debt. As a result of debt, many graduates delay getting married, having kids, and buying a house.[301] The trickle-down impact of improved household formation will mean higher demand for carpets, appliances, furniture, TV, and all other household items, which will create jobs.

Consumer spending accounts for 70 percent of the United States economy.[302] Bring forty million American consumers fully into the United States economy, and there will be more demand for cars, entertainment, travel, eating out, and other discretionary spending that will keep the United States economy growing.

Sixteen

United States Tax System

The changing demographics and advances in technology will serve as the catalyst for a transformation of the one-hundred-year-old tax code. Wealth and after-tax income inequality could shrink by focusing on reducing the cost of healthcare, education, and other factors that burden the lower and middle class. Reducing burden would increase their purchasing power. All of these benefits could be amplified with a fair tax code.

The current tax system penalizes working income and rewards nonworking income. The main goal of the current tax system appears to increase federal revenue to fulfill spending promises made by Washington. The government picks the winners and losers by determining tax rates and deductions. It spends according to political realities, and borrows to cover any shortfall in revenue. This system does not serve citizens who appear to spend more time serving the government instead of the government serving them.

Serving the Servants

In 2013, Americans worked 107 out of 365 days in a year just to pay federal, state and local taxes compared to 22 days 100 years ago when the tax system started (Fig. 16.1).[303] Americans spend almost sixteen weeks out of a year serving the government—(in order to pay taxes) that is supposed to work for them. To put it in perspective, Americans work almost four months to meet government spending needs before they can start keeping the money they earn for themselves.[304]

The federal tax code has ballooned to 70,000 pages as of 2010 from 400 pages when it started back in 1913. It becomes more complicated as it gets reformed. Reforms create more regulation, and new regulations open more loopholes for those with influence. They create more income for the tax compliance industry that generated revenues of $10 billion in 2014.[305] The tax-compliance industry is six times larger than the auto industry.[306] In 1935, Form 1040 had 34 lines and the instruction booklet had two pages, which has mushroomed to 79 lines and 209 pages by 2014.[307]

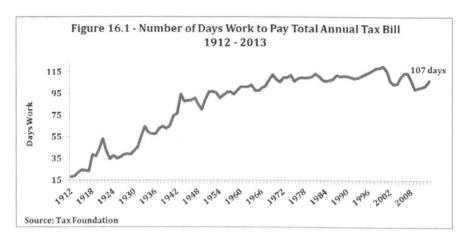

Figure 16.1 - Number of Days Work to Pay Total Annual Tax Bill 1912 - 2013

Source: Tax Foundation

American taxpayers spend about 6.1 billion hours a year[308] to comply with the federal income tax; that is equivalent to 3 million people working full-time year around who could have used their time more productively. According to the National Taxpayers Union Foundation, complying with the federal income tax cost the economy $233.8 billion in 2014 due to productivity loss.[309] Individuals and businesses as taxpayers must pay more than $1 in order to receive $1 of service or benefits from the government.[310] For every dollar they pay in taxes, they have to spend time in recordkeeping, organizing bills, buying software if they are self-filers or hiring tax preparers, all of which add up in more costs to taxpayers.

The growing on-demand and sharing economy allows earners, so called 1099ers, to serve as their own boss. The flexible workforce environment will change the income composition of Americans from a regular paycheck with a relatively permanent income to pay-for-performance with very few or no benefits. These earners pay the additional 7.65 percent in Social Security and Medicare tax that employers pay for their employees. Furthermore, they pay for health insurance. These emerging groups of workers will call for transparency and accountability from the government as they realize their overall tax burden.

First, Americans must recognize that the government does not make money by working or producing anything; it takes it from one taxpayer to give to another or keeps it. This approach can increase income inequality. Politicians get campaign contributions from one-percenters and corporations in return for influencing laws. On the other hand, the poor may end up becoming dependent on the government.

Second, Americans must acknowledge that the tax code is biased against the working income and is in favor of the nonworking income. Income from work is taxed at a higher rate than those who receive non-working income, known as passive income from investments. In addition, passive income earners do not pay Social Security or Medicare tax. Passive income is taxed at up to 20 percent and isn't subject to Social Security and Medicare tax (7.65 percent), but the working income is taxed at up to 39 percent plus 7.65 percent in payroll tax, more than double the nonworking income.

This favors wealthy Americans and corporations who will pay as little as legally allowed under the law. They used to pay up to 90 percent in marginal income tax during the 1960s and the economy did just fine.[311]

Politicians can get reelected by giving handouts to their constituents through tax codes, grants, and subsidies; most of the beneficiaries are their campaign donors such as corporations and the one-percenters. That money comes at the expense of those who do not have the influence in Washington. These incentives have increased income and wealth inequality in America at the expense of the middle class, despite all the tax reforms over the years.

There seems to be unanimous agreement that the tax code needs to be simplified, but the question is *how*? Politicians respond to their constituency by bringing federal money to their district in order to get reelected. However, they do this by adding provisions in the laws, including tax codes, known as tax expenditures, to benefit their constituents. These tax expenditures include tax credits, accelerated depreciation for businesses, mortgage deductions for homeowners, and more. Once established, it is almost impossible to eliminate them and they normally become a permanent part of the budget.

The only way to fix the complicated tax code and the inequity it causes is to implement transformational solutions and improve the political process. One of the concepts could be the decentralization of the taxing process based on the value of services provided by the local, state and federal government. That idea is to make the tax system simple and efficient as proposed through the Value Based Tax (VBT) system.

Value-Based Tax (VBT)

A value based tax (VBT) system would be a complete transformation of the total tax system. Conceptually, it would be analogous to the pay-for-performance model as discussed in earlier chapters. The tax rate pyramid could be distributed with the highest tax rates at the local level and the lowest tax rate at the federal level—the opposite of the current structure. For example, an individual living in California paid up to 39 percent tax rate at the federal level versus 12.3 percent at the state level in 2015.

Americans get the most government services at the local level and the least at the federal level. Therefore, it makes sense to have a higher tax rate at the local level and the lowest rate at the federal level. In return, the Federal Government will transfer most of the domestic responsibilities to state and local governments and focus on issues vital to the country such as defense, foreign policy, minority rights, and currency. This would hold true to the decentralized federal system envisioned by the country's founders and would bring more transparency and accountability to the way government spends money at all levels.

The basic premise of VBT is that the government is here to serve the people, and that people should pay a fair value for those services—but no more. The market can determine the value of services provided by the local, state and federal government, and how much taxpayers should be charged for those services. This approach is much more efficient than the current mode of spending first and then raising cash to pay for it.

The Federal Government generates its income through federal taxes, some of which are distributed to the state and local governments in the form of grants. Of course, those grants come with strings attached, some loosely and some very stringent. They generally fall into two main categories: block grants and categorical grants.

Block grants are money sent to state and local governments with more flexibility on how to use them. They are general purpose grants for local law enforcement, community development, social services, and Temporary Assistance for Needy Families (TANF). Categorical grants are a lot more restrictive and have a narrowly defined purpose. They could be a Head Start program for early childhood education, Special Nutrition Program for Women, Infants and Children, (WIC), and other similar federal initiatives.

The federal budget for fiscal year 2015 provided $641 billion in federal grants to state and local governments, or about 16.4 percent of the total United States budget, compared to $7 billion, or 7.6 percent, in 1960, according to the OMB (Fig.16.2).[312] The Federal Governments' control through the tax code has been growing since 1960. President Nixon argued that federal aid was a "terrible tangle" of overlap and inefficiency. In his 1971 State of the Union address, he lambasted, "The idea that a bureaucratic elite in Washington knows best what is best for people everywhere," and said that he wanted to, "reverse the flow of power and resources from the states and communities to Washington."[313]

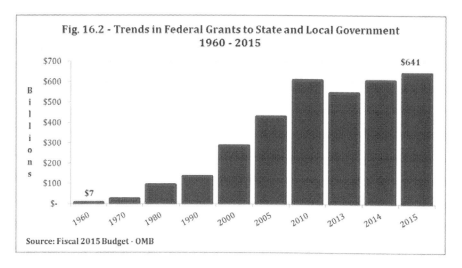

Fig. 16.2 - Trends in Federal Grants to State and Local Government 1960 - 2015

Source: Fiscal 2015 Budget - OMB

With VBT, the Federal Government can reduce federal taxes by up to $641 billion, or 16.4 percent, by not giving any of this money it collects from the taxpayers to states or local governments. Simultaneously, the federal tax rate could be cut in proportion. In return, it can allow the state and local governments to increase or decrease their tax rate according to their need. This will bring transparency and accountability closer to home.

Seventeen

Personal Taxes

A transformed tax system could place $4 billion of extra spending money in the middle-class families' pocket, in time saved. As technology evolves and 85 million millennials challenge the current system, Americans may no longer have to dread April 15 or spend energy collecting and organizing receipts, calculating earnings and deductions.

The last chapter discussed how different tax rates for working and nonworking income increase disparity. That was highlighted by Billionaire Warren Buffet in 2012 when he stated that his secretary pays a higher tax rate (35.8 percent) than he does (17.4 percent). The reason is that almost all of his income was from investments, which was then capped at a 15 percent capital gain tax rate that later was increased to 20 percent starting in 2013.[314]

Americans are being taxed everywhere they turn. In California, for example, residents' cable and phone bills include FCC fees, franchise fees, federal taxes, and utility user fees among others. Moreover, up to 18 percent of a wireless bill counts for federal, state, and local fees and taxes.[315] Then there is sales tax in most of the states that can be as high as 10 percent. A Californian's property tax bill has additional taxes, including a community college tax, a storm drain assessment, and a fee for trauma and emergency services. For every gallon of gas bought, almost 48-cents went to state and federal taxes as of March 2016.[316] Now there is the ACA tax disguised as a penalty if one is not enrolled in a health insurance plan.

The millennials, one of the largest voting blocs, who grew up with very complex devices, yet very simple to use, such as tablets and smartphones, may demand simplicity. As they move up the career ladder and make more money—spending more and more time and money on taxes and fees—they will have more incentive to call for a transformation of the tax system. Just as consumers revolted against Bank of America's debt card fee or Netflix's increase in subscription rate, millennials are likely to share their verdict in the upcoming elections by supporting candidates who call for the transformation, simplification, and decentralization of the tax system, as discussed in the Chapter Twenty.

Personal taxes take a significant amount of the middle-class's paycheck. For example, if one earned less than $118,500 in 2015, he or she paid 7.65 percent in payroll taxes, which plummets to 1.45 percent for any income over that amount. Thus, for incomes higher than $118,500, federal taxes drop by $6,200 per $100,000 above that threshold.[317] In addition, low or middle-income families spend a higher percentage of their earnings on day-to-day purchases that are hit by sales tax (in most of the states).

Disparity is more prominent between working, nonworking and corporate taxes as shown in Fig. 17.1 for $50,000 in taxable income.[318] According to a 2013 report by the Institute on Taxation and Economic Policy (ITEP), "Combining all of the state and local income, property, sales and excise taxes state residents pay, the average overall effective tax rates by income group nationwide are 10.9 percent for the bottom 20 percent, 9.4 percent for the middle 20 percent and 5.4 percent for the top one-percenters."[319] Add federal taxes, and the middle class ends up paying more than 50 percent of their income in taxes and fees between all branches of American government.

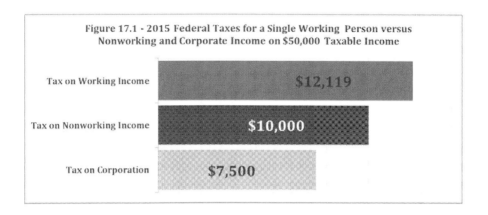

Figure 17.1 - 2015 Federal Taxes for a Single Working Person versus Nonworking and Corporate Income on $50,000 Taxable Income

Tax on Working Income	$12,119
Tax on Nonworking Income	$10,000
Tax on Corporation	$7,500

Meanwhile, personal income tax as a percentage of the total federal income has remained around 45 percent for decades, but the corporate share has declined from 40 percent in the 1940s to 13 percent in 2014 (Fig. 17.2). The gap between taxes paid by individuals and corporations has been widening at a faster pace since the 1960s.

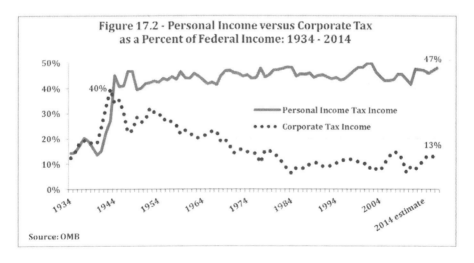

Figure 17.2 - Personal Income versus Corporate Tax as a Percent of Federal Income: 1934 - 2014

Source: OMB

The solutions range in difficulty due to the political process, but they all could be implemented over a five to ten-year horizon with the right narrative and leadership.

Eliminate Tax Deductions

In 2011, the Federal Government collected $1.1 trillion from individual taxes, but recorded $1.1 trillion in tax deductions, also known as expenditures.[320] It would have collected $2.2 trillion if tax deductions were eliminated. The government could have cut the tax rates by half across the board and would still have collected the same amount.

The current tax system benefits less than one-third of taxpayers who use tax deductions.[321] According to the IRS data, almost 70 percent of the taxpayers take basic standard deductions when they file their tax returns. That rate is very consistent dating back to 1990 as shown in Fig. 17.3. Major tax expenditures include interest on mortgage loans and exclusion of employer-sponsored health insurance from income. Millennials are less inclined to own a house, so the home mortgage tax deduction would have less value to them. The argument against the elimination of tax expenditures is that it will be disruptive to the economy, but it will benefit two-thirds of the taxpayers who would not have any tax deductions.

One way to put more money in low and middle-class pockets is not to file any tax return but to get a check or a bill from the Federal Government. This would not affect federal tax income, but it would save taxpayers 500 million hours that could be used for more productive work or time spent with family and friends. It will not require any changes in tax law. The only thing to do is to change the tax reporting process for 101 million tax filers who file simplified tax returns.[322] Moreover, the IRS will be able to significantly reduce its operational cost since two-thirds of the tax returns will be automated.

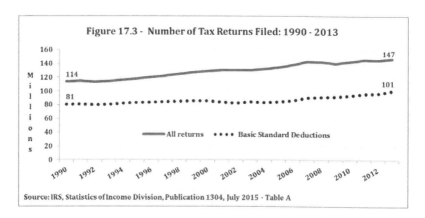

Figure 17.3 - Number of Tax Returns Filed: 1990 - 2013

Source: IRS, Statistics of Income Division, Publication 1304, July 2015 - Table A

In 2013, 101 million out 147 million taxpayers filed simplified tax returns with no deductions (Fig. 17.3).[323] According to the IRS disclosure, the average cost burden for a simplified tax return is $40 per return, or $4 billion, that the lower or middle-class family can save since they are the ones who would file simplified returns and potentially do not have any tax deductions to take.

Here is how it could work: the government collects individual wages and tax withholding information from employers through the W-2 forms. It also collects interest and dividend income from financial institutions and any payments to independent contractors through the 1099 forms. Therefore, it has all the basic income information. The IRS can easily analyze those 100 million taxpayers' tax status. If they owe money or are supposed to get a refund, the IRS can send them either a bill or a check.

This idea has been proposed in some form or another at the state level starting with California. State Controller Steve Westly, a former eBay executive, made his mission to use technology to make the state system user-friendly. The pilot program known as ReadyReturn started in 2005.[324] Under this program the state prefills the information and sends the return to the taxpayer to review and sign. Other countries such as Denmark, Chile, and New Zealand have similar programs. There is no reason the IRS cannot do the same for basic simple tax returns at the federal level.

Simplified Progressive Tax

Eliminating payroll tax by including it as part of the federal tax would benefit the middle class. Known as the Federal Insurance Contribution Act (FICA), it is imposed on the employee and employer to fund Social Security and Medicare—but it really is not the insurance it proposes to be. Social Security collects 12.4 percent of the earnings combined from the employer and the employee. However, it is capped at earnings of $118,500 in 2015. Taxpayers earning more than $118,500 do not pay any FICA tax above that amount. In addition, there is a Medicare tax of 2.9 percent without any cap. These two taxes amount to 15.3 percent in total cost per dollar of wages that has a disproportionately negative impact on the lower and middle class. Making them part of the general federal tax makes things simpler and could help reduce income inequality.

The Congressional Budget Office (CBO) that provides non-partisan analysis to Congress breaks down the American household into five quintiles based on income. According to the CBO report of November 2014, the average individual tax rates were 11.2 percent, 15.2 percent, and 23.4 percent for the middle, fourth, and the highest quintiles.[325] However, without payroll and other taxes, it goes down to 2.4 percent, 5.8 percent, and 14.2 percent respectively.

There is no question that combining the payroll tax and eliminating tax deductions will politically be a tough sell, but it would be fair and will take the government out of picking and choosing winners and losers through its tax policy. This simple system could be more efficient by implementing it in conjunction with three tax brackets: 5 percent, 15 percent, and 25 percent for the third, fourth, and fifth quintiles respectively, or a similar distribution calibrated not to add to the deficit. This proposal would simplify the system and shift the burden away from the middle class.

Eighteen

CORPORATE TAXES

America's corporate tax structure, like the personal tax structure, dispro-portionately drains the middle class. Transforming corporate taxes will help bring jobs back to the United State, make corporations more competitive in the global market, and more revenue to the United States Treasury. Some of the other weaknesses of the corporate tax code, such as tax expenditures (tax deductions), offshore income, inversions or the executive clawback provisions, are under public scrutiny and are being addressed to bring some fiscal stability to economic policy. Income disparity between workers and corporate executives and lower tax rate or no tax paid by corporations affect the middle class directly.

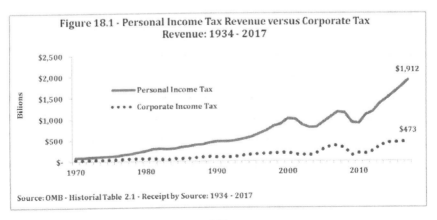

Figure 18.1 - Personal Income Tax Revenue versus Corporate Tax Revenue: 1934 - 2017

Source: OMB - Historial Table 2.1 - Receipt by Source: 1934 - 2017

Bashing corporations has not produced any results. If anything, it had a negative impact as corporations' share of the federal income has been declining over the years as shown in Fig. 18.1.[326] Large corporations can afford high paid lobbyists to make sure that their interests are protected and that they pay the lowest possible taxes. In the 1970s, the Federal Government's revenue from corporate and personal income tax was about the same. However, corporations' contribution has declined to about one-fourth that of personal taxes by 2017.

Two Sets of Accounting Books

Corporations keep two sets of books, one for the IRS and the second for the shareholders.[327] They show lower taxable income for IRS purposes and the highest possible income to their shareholders to keep their jobs and bonuses and also to maximize their stock's value. They pay lower taxes by reporting lower income for tax purposes even though they would make more money according to the generally accepted accounting principles known as GAAP. Corporations use GAAP accounting to report income to their shareholders.

Corporations spent $110 billion dollars in 2009 for tax compliance according to an IRS report[328]—which has likely risen since that time. This expense only adds to the cost of the goods or services produced by the corporations with the ultimate burden falling on the consumers.

By simply eliminating two sets of books and taxing the corporations based on what they report to shareholders, the IRS could significantly increase the federal tax revenue. Part of the additional income to federal revenue can be transferred back in a reduced corporate tax rate for small businesses. Adding that to the reduced corporate tax compliance expense will make United States corporations very competitive and more profitable in the global economy.

Tax Expenditures

Similar to individual tax deductions, eliminating corporate tax expenditures could double federal revenue or allow the corporate tax rate to be cut in half without impacting the current collection amount. Estimated revenue losses in 2011 were about $181 billion, the same amount that the federal government collected in that year, according to a GAO report.[329] The government could proportionately reduce the budget deficit by eliminating all tax expenditures over a ten-year period to minimize any disruption in the economy.

Eliminating all tax expenditures would increase federal tax revenue and also reduce the tax compliance cost to corporations. It will make corporations more efficient. Furthermore, investment decisions will be made on their merits without the influence of the corporate tax code. On the other hand, some corporate tax expenditures are particularly detrimental to lower and middle-class Americans like offshore income.

Offshore Income

American multinational corporations often keep their money overseas and invest overseas, while foreign companies are establishing plants in the United States. One possible reason is the United States corporate tax code. Corporations have increased their offshore profits that are parked overseas from $434 billion in 2005 to $2.1 trillion as of September 2015, according to Americans for Tax Fairness. More than half of those profits can be attributed to the high tech and healthcare industry, the one that traditionally provides higher paying jobs.[330] If they bring those profits home, they would be paying a 35 percent tax rate. However, they do not pay any taxes as long as they keep those profits overseas.

This is one of the major flaws in the corporate tax code. Corporations are incentivized to keep their profits in foreign countries, make investments, and create jobs in those countries rather than in the United States. By changing the current approach, the United States could create high paying jobs for the middle class and increase tax revenue. Even imposing a lower tax rate of 14 percent on offshore profits, as President Obama sought in early 2015, would have brought an additional $294 billion to the United States Treasury, but that effort failed.

The counter-productive corporate tax system was illustrated when Apple issued $17 billion in corporate bonds even though it had $145 billion in cash on its books in 2013, $100 billion of which were held in foreign bank accounts. Corporate bonds are issued by them to raise funds that are tax deductible for corporations. If they transferred $26 billion ($17 billion plus $9 billion tax) of that $100 billion to the United States, the company would have paid as much as $9 billion in federal taxes.[331] Therefore, it was cheaper for them to raise the debt in the United States instead of paying $9 billion in taxes. Furthermore, they were able to take tax deductions from the interest paid on that debt. Not only did the Federal Government lose that potential income, but they also continued to lose money on reduced profits for Apple due to the interest cost on that debt.

Inversion

Inversion is a growing phenomenon where United States-based companies change the address of their headquarters overseas and incorporate them in a foreign country. This method results in the elimination of offshore profits for tax purposes. Domestic corporations are saddled with debt from their overseas company, which reduces the income of their domestic corporation since the interest on debt is tax deductible (another argument to eliminate the tax deductions). At the same time, the overseas company will show income from the interest on that debt. Since these corporations are headquartered overseas in tax haven countries, with little or no tax on corporations, they pay very little if any local taxes. Tax haven countries are known to have very low or no tax on corporations. Therefore, those corporations avoid paying any tax on their income by shuffling papers between the U. S. and their foreign headquartered companies. At least fifteen publicly traded companies have taken steps over the last three years to incorporate abroad. Most of their CEOs did not leave the country but their profits did.[332]

As Congress started a crackdown on this strategy, these corporations are now shifting their focus to acquiring foreign companies. They set the new merged company in such a way so that the foreign companies own a bit more than 20 percent of the newly merged company. Pfizer, one of the largest United States drug makers, in November 2015 announced its plan to merge with Allergan, domiciled in Ireland, to save $1.7 billion in taxes through inversion in 2018.[333] It ended up terminating the merger due to the government's pressure.

Carried Interest

Carried interest is particularly beneficial to Wall Street—private equity and hedge fund managers, who pay only 20 percent federal tax on all of their income. They are primarily investment management companies that cater to wealthy individuals, pension funds and other large entities. Normally, one must have at least $1 million in net worth before these companies will take the investor as a client.

According to then-Congressional Budget Office Director Peter Orszag's July 2007 testimony before the Senate, private equity and hedge funds managed a total of $2 trillion combined. The top twenty-five hedge fund managers earned $14 billion in 2006, with one manager earning more than $1.5 billion.[334] This means that Wall Street had $2 trillion in funds to buy and sell companies, cut payrolls and costs, employ financial engineering techniques, and sell them to new investors at huge profits that are taxed at maximum of 20 percent.[335]

Simple Solution

Various types of corporate tax deductions continue to help large corporations avoid paying any tax or pay as little as possible. The lower payment of taxes by corporations transfers the burden to the middle class in the form of higher taxes to them or debt on their children since federal spending continues to rise. A fair and simple tax code would make United States Corporations more productive and competitive where they will have incentives to make investment decisions based on economics without having to look at the tax consequences of those decisions.

The United States' corporate tax policy is geared to encourage investments through tax credits and deductions to create jobs.[336] However, the United States economy is consumption-based—household purchases account for 70 percent of the economy.[337] Local, state and Federal Governments have been providing all kinds of tax incentives to corporations for investing in the United States; instead corporations appear to be buying back their stocks. In most recent years, share buybacks reached a record $520 billion, the money that could have been used for investment supposedly to create jobs.[338] All the while, the economy is faltering or growing very slowly despite billions of dollars in financial stimulation by the government since the financial crisis of 2007-2008.

If consumers have no money to spend, they limit their purchasing. As a result, there is no demand for products, so businesses don't build new factories or invest. Therefore, it only makes sense to simplify the corporate tax code. The simple solution is to eliminate two sets of books, saving millions of dollars in compliance costs, and eliminate all tax deductions. Making these changes over a ten-year period would minimize the disruptions of corporate operations and planning.

All these policies, whether healthcare, education or tax, lead to the United States Congress, chartered to make laws, and the White House that approves or vetoes those laws. They both reside in the Beltway and are responsible for making or destroying the middle class.

Nineteen

The American political process is broken and neither one of the two major parties, Democratic nor Republican, seem to have any desire to work together. Only 9 percent of likely United States voters think that Congress is doing a good or excellent job.[339] Donald Trump has broken all political norms and protocols by challenging the establishment and winning the election. Yet, he had to run under the Republican Party's name to get there without the support of the party leaders.

The Beltway is the biggest impediment to economic growth and the main obstacle to the middle class comeback. Former Vice President Hubert Humphrey described the Beltway as, "26 square miles of Washington D.C. surrounded by reality."[340] Larry Fink, the CEO of BlackRock, the biggest asset manager on the planet with $4.6 trillion of clients' fund under management stated that the economy would grow by 1 to 1.5 percent more than what it did in 2015 "if we had a Washington that works."[341] That incremental growth would have created millions of new jobs for the middle class.

In reality, the Beltway has morphed into a beast, and one can call it The Beltway Beast.[342] The Beltway consists of Washington, D.C., and surrounding counties, and the Beast includes defense contractors, lobbyists, Wall Street, think tanks, advocacy groups, journalists, foreign agents, and everyone else who wants something from the government. All of these congregate in the Beltway and are actively engaged in soliciting legislative favors and taxpayers' money. Then there is Congress and the White House: both need lobbyists and Wall Street for campaign funds. All of these groups are living in a bubble away from reality that makes the Beltway Beast.

Rather than blame the one-percenters for the problems of the middle class, the issues come straight from Washington. "Right now the biggest threat to our national security, as far as I am concerned, is the paralysis in Washington," said Robert Gates, the former CIA chief and the Secretary of Defense under President George W. Bush and President Obama.[343] In the words of Republican Sen. Tom Coburn of Oklahoma, today's Washington has become a "permanent feudal class," a massive, self-sustaining entity that sucks people in, nurtures addiction to its spoils, and imposes a peculiar psychology on big fish and minnows alike.[344]

The government policies regarding healthcare, education, housing and taxes, discussed throughout this book, most of the time have been and continue to be counterproductive for the middle class. Every election cycle, politicians make promises to reduce healthcare costs, make higher education cheaper, housing affordable and taxes that are fair to lift the middle class. The results over the years have been higher healthcare, education, and housing costs, leaving the number of people living in poverty stagnant at 45 million[345] and a shrinking middle class.

Beyond these harmful policies, common Beltway practices, including earmarks, gerrymandering, and political rhetoric, have contributed to the breakdown of the political process and the middle class.

Earmarks

House Republicans established a self-imposed ban on the practice of pork-barrel spending through the earmark process in 2009. Pork-barrel spending is a metaphor for appropriating federal funds for local projects to a specific congressperson's district. In January 2011, they expanded the earmark moratorium for the entire House.[346] This moratorium remained in place as of 2015. A self-imposed moratorium is a significant improvement toward curtailing the abuse of power and the process by the elected leaders from both major parties.

One of the ways that Senators and Congresspersons get reelected is by bringing money from Washington to their districts and states through earmarks. Earmarks are legislative provisions that set aside funds for a specific program, project, activity, institution, or location. They normally circumvent merit-based or competitive allocation processes and appear in spending, authorization, tax, and tariff bills.[347] Even though earmarks are a very small percentage of the federal budget, they are a symbol of abuse by the elected members.[348]

The most famous case of earmarks was a widely-publicized proposal known as "Bridge to Nowhere," by Senator Ted Stevens of Alaska in 2005. It would have connected the town of Ketchikan (population of 8,900) with its airport on the Island of Gravina (population of 50) at a cost to federal taxpayers of $320 million, by way of three separate earmarks in the highway bill. A ferry service runs to the island, but some in the town complain about its wait (15 to 30 minutes) and fee ($6 per car).[349] Even though the bridge did not get built, Alaska kept the funds and Alaskans did build a three-mile road to nowhere.[350] Another example is half a billion dollars in taxpayer money was spent during 2011-12 to build Abrams tanks that the army did not want, but the congressperson who represented the district where they are built wanted these tanks and was able to appropriate the funds.[351]

Gerrymandering

Every 10 years, based on updated census data, electoral districts' boundaries are redrawn in accordance with population change. In compliance with federal law, Congressional districts are designed by the states. In most states, the legislatures—controlled by the two major parties—draw the redistricting plans. Both major political parties try to draw the district boundaries to include areas that are favorable to their candidates, mostly incumbents, while excluding areas that are unfavorable. This is known as gerrymandering. Both parties have exploited their power in state legislatures to redraw Congressional districts to protect incumbents and minimize competition.

California passed a ballot initiative in 2008 that established an independent citizens redistricting commission to draw Congressional and state legislative district lines. It has fourteen members: five Democrats, five Republicans, and four nonpartisans. Ohio became the latest state to vote down gerrymandering by a vote of 71 to 29 percent in November 2015,[352] which could encourage similar efforts in other states. Even the Supreme Court ruled in 2015 that the state can try to remove partisan politics from the process of drawing political maps.[353]

In December 2015, the Florida Supreme Court replaced the Republican-drawn Congressional districts that favored Republican candidates. The court stated that Republican lawmakers violated a 2010 constitutional amendment that prohibited legislatures from drawing districts that favor incumbents or favor one party over another.[354] This is a major victory for the American public in removing the two major parties' monopoly in electing leaders.

Engagement versus Isolationism

The Beltway's definition of engagement in international issues seems to be military presence or intervention. Even small world events are presented as a threat to national security, even though America is the most secure nation geographically and by any other measure. If a leader is not in support of military intervention, then they are called an isolationist, which can carry implications of weakness even though the rest of the world knows that the United States is the strongest nation in the world.

Many consider United States leaders "patriotic" when they send soldiers into a war while those who oppose military action in regional or civil wars are considered "unpatriotic" or "weak." Entering into international conflicts endangers American lives, and it costs money. American military efforts in Iraq and Afghanistan has cost American taxpayers as high as $6 trillion,[355] with very little or no benefit to the American, Afghani, or Iraqi people.

The American public is finally catching up with Washington and its symbiotic relationship with the military-industrial complex. The nation needs a new definition of engagement. In his 2016 State of Union, President Obama stated, "The United States of America is the most powerful nation on Earth—period. Period. ... We spend more on our military than the next eight nations combined. Our troops are the finest fighting force in the history of the world. No nation attacks us directly or our allies because they know that's the path to ruin. ... We also can't try to take over and rebuild every country that falls into crisis—even if it's done with the best of intentions. That's not leadership; that's a recipe for quagmire, spilling American blood and treasure that ultimately will weaken us. It's the lesson of Vietnam. It's the lesson of Iraq—and we should have learned it by now." He further described a new view of engagement using soft power, "The way United States military, doctors, development workers helped stop Ebola in West Africa."[356]

The Ghost of Past

America's leaders represent the World War II generation whose memory goes back to Hitler and Communism, but recent history shows that no super power or even nuclear power has been able to conquer and occupy other people without any consequences. Instead of fighting other people's wars, the American public wants to rebuild America first.

If there is any lesson to be learned from the past few decades, it is that venturing and occupying other nations is an existential threat to an occupying nation itself. The Soviet Union disintegrated into a number of nations and one of the reasons can be attributed to their occupation of Afghanistan in the 1970s. Now that the Soviet Union no longer exists and Communist China is America's banker, Washington found another ghost in the Middle East. The nation ended up in Afghanistan and Iraq and cannot seem to figure its way out of those countries. The military tried that in Vietnam to fight communism and lost thousands of Americans and millions of Vietnamese in the process.

The Beast knows that Afghanistan and Iraq are unwinnable wars,[357] just as former Secretary Henry Kissinger realized by 1966 that the U.S.

intervention in defense of South Vietnam was a doomed enterprise and that only a diplomatic solution would end the conflict. His private papers and diaries state:

"I am quite convinced that too much planning in the government and a great deal of military planning assumes that the opponent is stupid and that he will fight the kind of war for which one is best prepared. However… the essence of guerrilla warfare is never to fight the kind of war your opponent expects. Having moved very many large units into Vietnam … we must not become prisoners now of a large-unit mentality. Otherwise, I think that we will face the problem of psychological exhaustion."[358]

President Kennedy confided that:

"We don't have a prayer of staying in Vietnam … Those people hate us. They are going to throw our asses out of there at almost any point. But I can't give up a piece of territory like that to Communists and get the American people to re-elect me."[359]

President Johnson lamented:

"I can't get out I can't finish it with what I have got. So what the hell do I do?" (Referring to Vietnam War)[360]

These kinds of wars drag on because leaders follow misguided ideologies, don't admit error, and aren't fully accountable to the American people. The lower and middle classes suffer the most from those decisions as they make up the majority of the US armed forces. Yet, they kept on sending American soldiers to die. The same message seems to reverberate with the American engagement in Afghanistan and Iraq. History will find that leaders are still living in the past when it comes to international affairs.

Challenging Washington's Control

States and cities are challenging Washington's power by charting their own course on issues that are important to their community. California has taken a lead in questioning the two major parties' stranglehold in terms of the election process. More than 200 bills aiming to challenge laws and regulations coming out of Washington by various states were introduced in 2015.[361]

Governor Jerry Brown of California, in October 2015, signed a measure that will allow all eligible Californians to be automatically registered to vote at the Department of Motor Vehicles (DMV) when they visit for new driver's license or a renewal. The measure is also known as Motor Voter Act that could potentially add 6.6 million Californians to the voting roster who have not registered to vote.[362] Eleven other states, including Colorado, Illinois, Utah, and West Virginia, have passed laws during 2013-14 to modernize their voter registration system to make it easier for eligible citizens to sign up online.[363]

California passed a law in 2014 known as the State Sanctuary Law that advises police not to arrest a person for violating the immigration code.[364] This is a clear challenge to Washington's authority, whether one agrees with the law or not. Three hundred forty cities are fighting the immigration enforcement laws by not prosecuting illegal aliens.

West Virginia, New Jersey, Texas and other seventeen states have introduced legislation to allow terminally ill patients to have access to investigational products that have not been approved by the FDA.[365] In addition, 36 bills in 2015 were introduced in eleven western states aimed at taking control of public lands away from the Federal Government. Six of the bills were passed in Nevada, Utah and Montana.[366]

Beginning in January 2011, California's Top Two Candidates Open Primary Act went into effect whereby the top two vote-getters in the primary election, regardless of party affiliation, move on to the general election. This law applies to most of the federal offices except for candidates running for United States President. Under this act, the top candidates could be from the same party as was the case in California's senatorial race in 2016 when Kamala Harris and Loretta Sanchez, both Democrats, ran against each other.

At the national level, some Republican members of Congress are trying to fundamentally alter the centralized power structure of the House by reducing the power of the speaker in favor of greater authority for rank-and-file. Some members feel that they do not have a voice at the table or power in terms of what comes to the floor. The proposed changes would include stripping the speaker's power over the Republican steering committee, which appoints the chairman for all committees and the appropriation subcommittees.[367]

Transforming the Political System

The American system is dynamic; Americans continually have the option to vote for change. It seems every few decades sweeping changes come that have the chance to clean the system. After the Vietnam War and the Watergate scandal, Americans elected a Washington outsider, a peanut farmer and former governor of Georgia, Jimmy Carter, to start anew. Thirty years later, after the Afghanistan and Iraq War, a first-time senator, Barack Obama, with no executive experience became the first African-American President of the United States. He offered hope, a changed agenda, and brought a significant number of younger Americans to engage with the political process. As Beltway outsiders, both President Carter and President Obama faced significant resistance in pursuing their agenda from Congress since it has the power of writing checks. The jury is out on how effective President Trump will be in challenging the Beltway Beast. Regardless of the outcome, there is a need for a paradigm shift to transform the political system from bottom up.

Political leaders themselves often call for a change. In his January 20, 1981 inaugural speech, President Reagan said, "Government is not the solution to our problem, government is the problem." He ended up with the largest expansion of Medicare in 1988,[368] lowered the marginal tax rate for the top 1 percent and increased debt at an unprecedented level at that time. Fast forward thirty years to President Obama, in his 2012 state of the union message said, "Washington is broken." He further stated, "I've talked tonight about the deficit of trust between Main Street and Wall Street, but the divide between this city and the rest of the country is at least as bad—and it seems to get worse every year." Unfortunately, not only nothing has changed in Washington, but it has gotten worse in the last thirty years. These are man-made problems, which means, they could be fixed.

The American electorate has sent a clear message by electing Donald Trump as the President of the United States as an anti-establishment candidate and an outsider. The Democrats and Republicans together have deliberately limited choices of political leaders offered to the American public. Both major parties are so focused on winning that they have forgotten to govern or serve American people, and that presents a great opportunity for the independent movement.

Twenty

In the current two-party political structure one-percenters have and will continue to get richer, and corporations will gain more influence and pay lower taxes. President Trump is presiding over the richest cabinet in history with billionaires and multi-millionaires with a net worth of $14 billion in aggregate.[369] However, the 2018 election and beyond could be the beginning of the political transformation in America through the power of the people. The ninety-nine–percenters have the power of vote and it is *free*. The independents represent the silent majority. There is a growing mood against the establishment as manifested during the 2016 presidential election. Both major parties have been engaged in an anti-competitive behavior in electing our leaders.

Dem-Rep Duopoly

"The two parties have effectively created a duopoly… It's a scandal, really, because democracy should be about giving people alternatives and allowing peoples' voices to be heard. But the two parties collude to make sure that you don't get a third party… If you look at America's two parties, they're actually very close together in terms of their ideological differences. Both American parties—the Democrats and the Republicans—would fit comfortably as center-right parties in Europe," wrote journalist Fareed Zakaria on his CNN's Global Public Square blog.

In the United States politics, the oft-repeated cliché is that Democrats represent higher taxes and bigger government, while Republicans are advocates for decentralized government and fiscal responsibility. These distinctions are largely the product of a myth. The Democratic Party was founded in the late 1700s as an advocate for de-centralized government with limited power. In the 1792 election, followers of Thomas Jefferson dubbed it the Republican Party. By 1798, it became known as the "Democratic-Republican party," eventually becoming the Democratic Party in 1850. However, Democratic and Republican parties are the same: both parties have been piling up debt on the American people while the middle class has been shrinking over the last thirty years regardless of the party in power. Moreover, they both have been busy dividing the country with fear tactics by using national security as a shield.

On the domestic front, President Jimmy Carter deregulated the U.S. airline companies in 1978. He deregulated the Savings and Loans Industry (S&L) that was partly responsible for the housing crisis in the eighties and eventual demise of the S&L industry. Clinton repealed the Glass-Steagall Act from 1930s that caused the financial and housing crisis of 2008, according to some experts. This Act was designed to allow commercial banks to take deposits and make loans and investment banks to underwrite and sell securities but they could not do both.[370] Both of the above actions would be considered Republican domain.

Republican President Ronald Reagan increased government spending to $1 trillion from $321 billion and the civilian workforce to three million during his presidency. President George W. Bush expanded the entitlement by signing the Medicare Prescription Drug Improvement, and Modernization Act of 2003 (MMA), commonly known as Medicare Part D, which made the most significant changes in Medicare since 1965, notably including prescription drugs benefits. However, it was not funded.

On foreign affairs both parties seem to have relied on the American military power and arm sales to resolve international conflict. The nation has been engaged in the Middle East for over thirty years with the same policy. Yet, that policy has not produced any positive results for the region except that America has gained more enemies, thus making it less secure.

Third Party Movement

Over the last 60 years, there have been only two relatively successful attempts to get an independent presidential candidate elected. John Anderson in 1980 and Ross Perot in 1992 ran as independents. Both candidacies were primarily personality-based. During the 2016 election season, there was a talk about a potential third-party run by a Republican candidate supported by the party elites when Donald Trump was the presumptive Republican presidential nominee.[371] Trump won the election as a Republican candidate but did not follow the party platform and challenged the party's leadership.

The Reform Party was founded by billionaire businessman Ross Perot in 1995, which grew out of his presidential campaign as an independent in 1992. He had one of the most successful candidacies in 1992 garnering about 19 percent of the vote. Notable candidates were Jesse Ventura, who was elected governor of Minnesota in 1998, presidential candidates Pat Buchanan in 2000 and Ralph Nader in 2004. However, this party never gained any traction as a major national party.

Americans Elect sought to challenge the two-party system by placing a nonpartisan presidential candidate on the ballot in all fifty states for the 2012 election. It was founded as a 501(c) (4) organization in 2010 with $20 million, mostly from unnamed donors. The idea was to get on the ballot first and then nominate candidates for the President and Vice President. Historically, in third-party movements, candidates came first and organizations followed. Americans Elect did the opposite, building the organization first and hoping to find the right presidential candidate. However, this model did not work in the 2012 election due to the lack of an electable candidate.

The Tea Party is a high-profile organization composed of hundreds of grassroots cells throughout the country without any centralized structure. This group is composed mostly of disenfranchised Republicans and Libertarians. This is probably the most successful model in recent times for affecting policies and laws, as their caucus had shown during the 2013 government shut down.

The House Freedom Caucus, which is made up of 36 members, included many veterans of the Tea Party.[372] They have been successful as a voting bloc in defying House leadership, regardless if one agrees with their ideology or tactics. In the fall of 2015, they caused turmoil for Republican leadership. Having forty more independent candidates will further reduce the power of leadership from the two major parties.

Seizing the Moment

According to the U.S. Constitution, the requirement for a congressional candidate is, "No Person shall be a Representative who shall not have attained to the Age of twenty-five Years, and been seven Years a Citizen of the United States, and who shall, when elected, be an inhabitant of that State in which he shall be chosen." It does not require a candidate to be affiliated with a party.

Americans have been conditioned to believe that the two major party system is the best for America. Although it has worked well most of the last century, it cannot continue under the current reality where both parties are busy dividing the country instead of uniting it. However, the hurdles created by Democrats and Republicans are enormous for a successful middle-of-the-road third-party to take hold.

The 2016 anti-establishment mood of the public and the election of Donald Trump offer the best chance for an independent movement to take hold. It could eventually result in a mainstream third party, such as the Centrist Party[373] or People's Party of America.[374] President Trump is considered an outsider even though he ran as a Republican Party presidential candidate. The main challenge for any congressional candidate is to get on the ballot in his or her state. Each state has its own complicated laws which can vary from no signature required by Florida to as many as 35,000 in Washington. However, most of the states require fewer than 5,000 signatures. In addition, some states require a filing fee. Also, they are required to register with the Federal Election Commission.[375]

There is no realistic chance for an independent presidential candidate to win in 2020 because the structure of the process is controlled by two major parties. However, there is a very good chance for independent Congressional candidates to get elected in 2018 and beyond. It would take only thirty to forty independent Congress members to start transforming the country, by stopping any bad laws from passing and slowly eliminating the two-party duopoly.

The Democrats and Republican spend enormous amounts of money and time during their primaries to get their parties' nomination. Moreover, they tend to move toward the extreme views of their respective parties to win the nomination. Then they pivot to the center of ideological divide during the general election to appeal to the independents, the silent majority. This is the best opportunity that independent candidates can seize upon—first by saving all the money and time bypassing the primary process and second by questioning the credibility of their opponents who would shift their rhetoric from extreme to the center during the general election.

The Silent Majority

Forty three percent of Americans identified as politically independent in 2014, compared to thirty percent Democrats and twenty-six percent Republicans. On average, independents have been in the majority over the last twenty-five years (Fig. 20.1),[376] yet they virtually have no representation in Congress. It is not surprising, that Washington's policies do not represent the majority of Americans since the country is governed by parties that most Americans don't fully support.

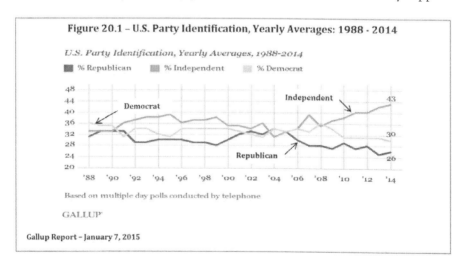

Figure 20.1 – U.S. Party Identification, Yearly Averages: 1988 - 2014

Gallup Report – January 7, 2015

In order to get elected for any federal office, both major parties Democrats and Republicans need independents' support since they both are in the minority. The alternative is for the independents to pick their own candidates, which is viable for Congress. An opportunity for changes to come is the demographics of the House and Senate. Congress is largely controlled by old men while almost 60 percent of the United States population is under the age of fifty, including one-third who are millennials. The average age of a Congressperson is 57 years and Senators 61 years in the 114th Congress that started in January 2015.[377] Twelve of the senators and twenty-two representatives are more than seventy-five years old, some even in their eighties.[378] As aging Congress members retire, it will create an opportunity for some new leadership and perspectives in the legislative branch—and as a result, this could start the process of transformation for the middle-class comeback.

Rebuilding the American Middle Class

Instead of promising the world as most of the politicians do, the independent candidates would have a better chance of winning the support of Democratic and Republican moderates alike in addition to independents. The following issues for independent candidates can serve as a platform for "Rebuilding the American Middle Class":

1. Reduce the interest rate on student loans to the same rates paid by Wall Street.
2. Set the term limit at three for Representatives and two for Senators.
3. Transform the tax system so that jobs stay in the country and move the power from Washington to the local and state level.
4. Oppose any military action unless Congress specifically declares war; it is paid for by a war tax and a draft is immediately implemented in order to defend the homeland.

These conditions are vital in transforming America.

The first pledge will address the forty million[379] Americans who are saddled with student loans that have impeded economic growth. The Federal Government owns 93 percent of the student loans.[380] The nation needs a pragmatic solution so that student loans are paid in order for the economy to grow at a faster pace. In the meantime, the Federal Government can reduce the burden by lowering the interest rate to what it charges big banks.

The second pledge will encourage average Americans to engage in the political process, secure in the knowledge that their elected leaders will not become a part of the Beltway Beast as career politicians. The Founding Fathers envisioned a system of citizen-politicians who would leave their jobs in times of a national crisis and join the government to lend their expertise and experience—and then return to private life once it is over. Americans must feel that they are the stakeholders and have leaders who are truly listening and working for them.

The third pledge is critical since the tax code is the main factor responsible for wealth inequality. It is also one of the reasons for the off-shoring of profits and jobs. A transformed tax system that is decentralized would shift the power away from the federal government to the state and local level and will bring the decision making closer to people.

The fourth pledge will require Congress to own up to its responsibility to declare war and be accountable for its decisions. In order to hold leaders accountable for the gravity of these decisions, this condition avoids the problems of recent wars. Americans weren't asked to pay for them nor were they required to send their sons and daughters to war. Americans would have never supported the war on Iraq if they had known that there were no WMDs and that it would cost thousands of lives and over $6,000,000,000,000 borrowed from their children. This will provide a necessary restraint to ensure that America only becomes militarily involved if the country is in an imminent danger, at which point every American will pitch in. This will also make sure that the composition of the military reflects the American people, as the lower and middle-class people are disproportionately represented.

The independent movement with the help of political disrupters could overcome the World War II generations' overreach and fear of others. It can help build and strengthen the middle class and bring Americans together.

Political Disrupters

Independent political candidates are essentially political entrepreneurs. With the help of technology, they can accelerate the pace of the transformation of the American political system. They want to build something better that would negate problems, such as gerrymandering or voter registration restrictions. Thirty-one states and the District of Columbia offer online voter registration.[381] This will allow more people, especially the millennials who are technologically savvy, to register and increase voter participation. Entrepreneurs are not afraid to learn from failures, and they don't place all of their faith in the opinions of others. They rely on themselves and their visions of a better world.[382]

Disruption occurs when entrepreneurs notice areas of weakness in the structure of ideas, institutions and incentives, and then find ways to change the institutional rules in those areas. They seek to fill the gap when they see a large problem and a lack of adequate solutions. In a democracy, this requires making sure favorable changes in laws are also in the political interests of the existing powers. Political entrepreneurs fulfill what John Stuart Mill said in 1845, "When the right idea and the right circumstances meet, the effect is seldom slow in manifesting itself."[383]

Disrupters have been successful in connecting the consumers with the suppliers directly without the so-called middleman. In the same vein, political entrepreneurs can shrink the long overdrawn primary process of nominating political leaders for elected offices. With the help of technology, they can reduce it to a one-day process through a hybrid version of physical and virtual conventions of political parties. The entrepreneurs may even be able to eliminate the need to physically going to the polling booths by bringing voting to the voters through smartphones or other electronic devices. Better yet, they may even open doors for a main stream third political party by creating a platform that bypasses the primary process altogether.

The election of President Trump as a disrupter may be the beginning of the political transformation in America. The time is right for political disrupters to transform the political system to make it more efficient, transparent, and open for all Americans to engage with the process of electing leaders—and ensuring the needs and goals of the middle class.

Twenty-One

It is a just matter of time before the American middle class comes back. Women are gaining a fair and equal share in private and public deliberation as the pace of progress accelerates. Millennials are re-defining American culture and values. Technology will serve as the catalyst for the middle class comeback through reductions in the cost of living due to productivity gains in critical factors affecting them. Those conditions offer an opportunity for disrupters to provide an affordable healthcare delivery model and an efficient education system that can put more money in the lower and middle-class families' pockets. If there ever was a time for the transformation of the education system, healthcare delivery model or political process, it is now.

The role of women, technology, disruptive platforms and thinkers across the country will certainly have an effect. The wild card in the middle class comeback is that the political system is broken. The Beltway Beast is the major impediment against the middle class comeback. This broken system that is very divisive is a major threat to American security as well. Well-intended promises and laws in healthcare, education, housing, or tax policies have hurt rather than help the middle class. The election of Donald Trump and his governing style and the unpredictability of his decision has added another wrinkle to the system. His trade policies of import tariffs could add further cost to the lower and middle class budget by making imported products more expensive. Most of the daily necessities are imported and sold through Walmart and other big box stores and could cost up to one trillion in higher costs to American consumers over the next ten years.

The middle class has been destroyed under the leadership of both parties over the last three decades. Their policies have given an unchecked rise to the cost of education and healthcare as well as taxes (directly or indirectly) while the income for the middle class has been about the same as it was in the 1980s.

Independent Congressional candidates are the hope of the middle class. They can form a caucus that can bring all the middle of the road pragmatic leaders from major parties and independents together to stop laws that hurt the middle class. At the end of the day, the President does not have the power to make laws but Congress does.

In an election, numbers reflect power. However, the majority 43 percent independents have no representation and are forced to choose between two major party candidates. The third party candidates are considered spoilers since the odds are stacked against them in winning the election. However, the 2018 election offers an opportune time for the independents to show their strength by electing independent Congressional candidates, which will eventually lead to a fiscally responsible and socially compassionate third party.

Winston Churchill once observed, "Americans will always do the right thing, only after they have tried everything else."[384] The 2016 election may be the watershed moment in United States history that will start the transformation of the political system and open the door for the middle class comeback led by women, millennials, and technology.

Acknowledgments

I see a great future for the American Middle Class as women, millennials, and technology pushes America to new frontiers. I have been very fortunate to have a loving and a compassionate mom, late Khatoon Moon, who saw nothing but goodness in people. I have been inspired by one of my millennial sons, Atif, who was born with cancer, had numerous surgeries and is wheelchair bound. He does not let his disability gets in his way and lives a successful life as a professional wheelchair tennis player and manages a small family business.

This book would not be complete without Atif's help in editing, advising and fixing my typos. Many thanks to Melissa Ann Wuske for helping keep the focus on the book's message while challenging me on the narrative and the tone of the book. I would also like to thank Carol Thompson for placing the final touches to the book, Shaireen for her critique and providing millennial's perspective and my son Jamal for the final editing. I am grateful to Lisa Hoffmaster, Andrea Cockrum, Professor Amir Hussain, Ajit Patel and Janine Anderson-Bays, for their comments and advice.

Finally, I would like thank my wife Elena for her unwavering support and guidance to work on this book.

End Notes

Introduction:

1. "Annual report of the White House task force on the middle class, The White House," Feb 2010 – Pg. 10

2. Angela Monaghan, "U.S. Wealth inequality – top 0.1% worth as much as the bottom 90%," The Guardian, Business, November 13, 2014

3. "Harnessing the Power of the Purse: Female Investors and Global Opportunities for Growth," Center for Talent Innovation, Executive Summary, 2014

4. "Millennials Outnumber Baby Boomers and Are Far More Diverse," Census Bureau Reports, June 25, 2015

5. Gasoline Tax, American Petroleum Institute, accessed April 2, 2016, http://www.api.org/oil-and-natural-gas-overview/industry-economics/fuel-taxes/gasoline-tax

6. Fact Sheet – Social Security, 2015 Social Security Changes, accessed March 14, 2016, https://www.ssa.gov/news/press/factsheets/colafacts2015.pdf

7. Jeanne Sahadi "Up to 35% of your cell phone bill may be taxes and fees," CNN Money, October 9, 2014

Chapter 1:

8. "Middle Class in America," Department of Commerce, January 2010, Page 1

9. "Middle Class in America," Department of Commerce, January 2010, Page 2

10. "Annual report of the White House task force on the middle class, The White House," Feb 2010 – Pg. 10

11. "The Distribution of Household Income and Federal Taxes, 2011," Congressional Budget Office, November 2014 (Table 1)

12. "Annual report of the White House task force on the middle class, The White House," Feb 2010 – Pg. 10

13. "The American Middle Class Is Losing Ground," Pew Research Center, December 9, 2015

14 "The Lost Decade of the Middle Class," Pew Research Center, Chapter 1, Aug 22, 2012

15 "Middle Class in America," Department of Commerce, January 2010, Page 1

16 "The Lost Decade of the Middle Class," Pew Research Center, Chapter 1, Aug 22, 2012

17 Jordan Weissmann, "Why Don't Young Americans Buy Cars," The Atlantic, March 25, 2012

18 Jim Harter and Sangeeta Agrawal, "Many Baby Boomers Reluctant to Retire," Gallup, January 20, 2014

19 "The Distribution of Household Income and Federal Taxes, 2011," Congressional Budget Office, November 2014, Fig. 8, Page 18

20 "The Distribution of Household Income and Federal Taxes, 2011," Congressional Budget Office, November 2014, Calculated from Table 1. Federal Taxes/Before Tax Income

21 "SEC Proposes Rules Requiring Companies to Adopt Clawback Policies on Executive Compensation" SEC Press Release 2015-136, July 1, 2015

22 "First on CNBC: Alibaba Founder & Executive Chairman Jack Ma Speaks with CNBC's Squawk on the Street," September 19, 2014, http://www.cnbc.com/id/102016827

23 Larry Platt, "Magic Johnson Builds an Empire," *The New York Times Magazine,* December 10, 2000

24 Tara Siegel Bernard, "In Retreat, Bank of America Cancels Debit Card Fee," *The New York Times,* November 1, 2011

25 Ibid

26 Greg Sandoval, "Netflix's lost year: The inside story of the price-hike train wreck," Cnet.com – July 11, 2012 http://www.cnet.com/news/netflixs-lost-year-the-inside-story-of-the-price-hike-train-wreck/

27 "The Sharing Economy," PWC.com, Consumer Intelligence Series (2015), 14

28 "The Sharing Economy," PWC.com, Consumer Intelligence Series (2015), 9

29 "15 Economic Facts about Millennials," The Council of Economic Advisers, October 2014, Page 11

30 Jordan Weissmann, "Why Don't Young Americans Buy Cars," The Atlantic, March 25, 2012

31 Martin Feldstein, "Opinion: Are middle-class incomes really stagnating," Market Watch, July 30, 2015

Chapter 2:

32 "Women in Labor Force," United States Department of Labor, Women's Bureau, accessed on March 7, 2015 http://www.dol.gov/wb/stats/facts_over_time.htm#earnings

33 "Earned degrees conferred by degree-granting institutions, by level of degree and sex of student: Selected years, 1869-70 to 2013-14," Department of Education – Digest of Education Statistics, Table 247 https://nces.ed.gov/programs/digest/d04/tables/dt04_247.asp

34 "Women in Labor Force," United States Department of Labor, Women's Bureau, accessed on March 7, 2015 http://www.dol.gov/wb/stats/facts_over_time.htm#earnings

35 Philip Cohen, "More Women Are Doctors and Lawyers Than Ever—but Progress Is Stalling," The Atlantic, December 11, 2012

36 "Women in the Labor Force: A Databook," BLS Reports, Bureau of Labor Statistics, December 2014, Table 25

37 Mark Hugo Lopez, "Women's college enrollment gains leave men behind," Pew Research Center, March 6, 2014

38 Caroline Fairchild, "Number of Fortune 500 women CEOs reaches historic high," Fortune, June 3, 2014

39 Mandi Woodruff, "Why Women just might save the middle class," Yahoo Finance, April 7, 2014

40 "Women's work – The economic mobility of women across generations," Pew Charitable Trust Report, Table 1, page 5, April 2014

41 "Women in the U.S. Congress 2017", Center for American Women and Politics (CAWP), accessed, February 12, 2017 http://www.cawp.rutgers.edu/women-us-congress-2017

42 Chris Cillizza, "18 million cracks and the power of making history," Washington Post - June 13, 2013

43 "Nancy Pelosi Fast Facts" CNN Library, accessed March 1, 2013 http://www.cnn.com/2013/03/01/us/nancy-pelosi-fast-facts/

44 Mollie Reilly, "The Women Of The Supreme Court Now Have The Badass Portrait They Deserve," Huffington Post, October 28, 2013

45 Philip Cohen, "More Women Are Doctors and Lawyers Than Ever—but Progress Is Stalling," The Atlantic, December 11, 2012

46 "Earned degrees conferred by degree-granting institutions, by level of degree and sex of student: Selected years, 1869-70 to 2013-14," Department of Education – Digest of Education Statistics, Table 247 https://nces.ed.gov/programs/digest/d04/tables/dt04_247.asp

47 "Women's work – The economic mobility of women across generations," Pew Charitable Trust Report, Table 1, page 5, April 2014

48 "Women's Earnings as a Percentage of Men's Earnings by Race and Hispanic Origin," U.S. Census - Historical Tables, Table P-40

49 Judith Warner, "The Women's Leadership Gap," Center for American Progress, page 2 – March 7, 2014

50 "Earned degrees conferred by degree-granting institutions, by level of degree and sex of student: Selected years, 1869-70 to 2013-14," Department of Education – Digest of Education Statistics, Table 247 https://nces.ed.gov/programs/digest/d04/tables/dt04_247.asp

51 Mark Hugo Lopez, "Women's college enrollment gains leave men behind," Pew Research Center, March 6, 2014

52 "Knowledge Center - Women in the United States," Catalyst, June 10, 2014, accessed December 30, 2015 http://www.catalyst.org/knowledge/women-united-states

53 Jennifer L. Lawless and Richard Fox, "Men Rule – The Continued Under-Representation of Women in U.S. Politics," Women & Politics Institute, American University, January 2012

54 Jennifer E. Manning, "Membership of the 114th Congress – A Profile," Congressional Research Service, January 22, 2015

55 Munir Moon, "Glass Ceiling in U.S Congress," Huffington Post Blog, September 18, 2014

56 Micah Zenko, "Walking Loudly and Carrying a Big Stick -Why women are less inclined to start wars," Council on Foreign Relations, August 6, 2013

57 "Minding the manufacturing gender gap" Deloitte and Manufacturing Institute, page 6, 2015

58 Bridget Bergin, "By The Numbers: The State Of Women In Manufacturing," Industrial Distribution, August 2015

59 Mandi Woodruff, "Why Women just might save the middle class," Yahoo Finance, April 7, 2014

60 Eileen Patten, "On Equal Pay Day, key facts about the gender pay gap," Pew Research Center, April 14, 2015

61 "Breadwinner Moms," Pew Research Survey, May 2013

62 Ibid

63 "Women in the Labor Force: A Databook," BLS Reports, Bureau of Labor Statistics, December 2014, Table 25

64 "Harnessing the Power of the Purse: Female Investors and Global Opportunities for Growth," Center for Talent Innovation, Executive Summary, 2014

65 "15 Economic Facts about Millennials," The Council of Economic Advisers, October 2014, Page 14

66 "The Economic Impact of Women-Owned Businesses In the United States," Center for Women's Business Research, Executive Summary, October 2009

67 "More Women Are Leaving Corporations To Become Entrepreneurs" Conference Board Press Release, March 27, 2003

68 "2014_State_of_Women-owned_Businesses Report" Commissioned by American Express Open, 2014

69 Chad Brooks, "12 Businesses You Didn't Know Were Started by Women" Business News Daily, July 20, 2015

70 Kiva - About Us - Viewed on December 3, 2015

Chapter 3:

71 "The future of mobility," Deloitte University Press (2015), 7

72 Ibid, 10

73 William Diem, "Experts See Autonomous-Car Revolution as Evolution," WardsAuto, March 5, 2014

74 Jordan Weissmann, "Why Don't Young Americans Buy Cars," The Atlantic, March 25, 2012

75 Brishen Rogers, "The Social Costs of Uber" University of Chicago Law Review Dialogue (2015), 85

76 "Uber Drive with Uber – Make Money on Your Schedule" accessed March 17, 2016 https://www.uber.com/drive/

77 "The future of mobility," Deloitte University Press (2015), 15

78 Adi Robertson, "Here are the three things Amazon needs to get its delivery drones off the ground," The Verge, December 2, 2013

79 Matthew Chambers, Mindy Liu, and Chip Moore, "Drunk Driving by the Numbers," Bureau of Transportation Statistics Publications, United States Department of Transportation April 2012

80 Daniel Thomas, "Driverless convoy - Will truckers lose out to software," BBC News, Business, May 26, 2015

81 "Crashes are the leading cause of on-the-job death for truck drivers in the US," Centers for Disease Control and Prevention, Press Release, March 3, 2015

82 Daniel Thomas, "Driverless convoy - Will truckers lose out to software," BBC News, Business, May 26, 2015

83 "Energy Cost Impacts on American Families," American Coalition for Clean Energy, (June 2015), 6

84 Patti Domm, "US energy is growing, and so is US 'power,'" CNBC, November 25, 2014

85 "Sustainable Energy in America," The Business Council for sustainable Energy, 2015 Factbook, Executive Summary, February 2015

86 Ibid

87 Jon Fingas, "Spray Painted solar cells promise cheap power on seemingly any surface," engadget, August 3, 2014

88 Stu Robarts, "The Paint on the Mercedes Vision G-Code concept harvests solar and wind energy," Gizmag, November 4, 2014

89 "V2G car generates electricity--and cash," University of Delaware Daily, November 2008

90 "Energy Consumption by Sector," U.S Energy Information Administration, Monthly Energy Review, December 2015, Table 2.1

91 "67% Electricity generated from fossil fuel" EIA, accessed October 25, 2015 http://www.eia.gov/tools/faqs/faq.cfm?id=427&t=3

92 "Sea level rise will swallow Miami, New Orleans, study finds," Phys.org – October 12, 2015

93 Tracey Lien, "Bending razor-thin glass to tech's future needs," LA Times, October 25, 2015

Chapter 4:

94 "The Millennial Generation Research Review," U.S. Chamber of Commerce Foundation (2012), 14

95 "The $30 Trillion Challenge," Morgan Stanley - Wealth Management, April 28, 2015

96 "Financial Marketer's Guide to Mass Affluent Millennials," Equifax – 2015

97 "15 Economic Facts about Millennials," The Council of Economic Advisers, October 2014, 3

98 "Comparing Millennials to Other Generations," Pew Research Center, March 19, 2015

99 "Millennials Outnumber Baby Boomers and Are Far More Diverse," Census Bureau Press Release (CB15-113), June 25, 2015

100 "Millennials in Adulthood," Pew Research Center, Social and Demographic Trends, March 7, 2014

101 "The Millennial Generation Research Review," U.S. Chamber of Commerce Foundation (2012), 11

102 "On Pay Gap, Millennial Women Near Parity – For Now," Pew Research, December 11, 2013, 2

103 Matt Kapko, "How Facebook Stimulates the Economy and Creates Millions of Jobs" CIO Magazine, January 27, 2015

104 Kurt Wagner, "8 Ways Facebook Changed the World," Mashable.com, February 4, 2014

105 Ibid

106 Mark Wilson, "How Facebook Just Became The World's Largest Publisher," FastCompany Magazine, May 13, 2015

107 Colleen DeBaise, "How Danae Ringelmann Built Indiegogo," The New York Times, December 11, 2014

108 Ibid

109 "Success Stories, Misfit Shine Changes the Game for Wearables," accessed March 17, 2016 https://learn.indiegogo.com/success/

110 Ben Sisario, "With a Tap of Taylor Swift's Fingers, Apple Retreated," The New York Times, June 22, 2015

111 Lyanne Alfaro, "Your $220 million to the ALS bucket challenge made a difference, study results shows," Business Insider, August 20, 2015

112 Julian Chokkattu, "Warby Parker Hits One Million Glasses Sold, Distributed," TechCrunch, June 25, 2014

113 Andrea Chang, "Mark Zuckerberg and Priscilla Chan pledge to donate 99% of their Facebook shares," LA Times, December 1, 2015

114 "Millennials in Adulthood," Pew Research Center, Social and Demographic Trends, March 7, 2014

115 "Millennials - The Politically Unclaimed Generation," The Reason-Rupe Spring 2014 Millennial Survey, 24

116 Ron Fournier, "The Outsiders: How can millennials change Washington if they hate it?," The Atlantic, August 26, 2013

117 John Eligon, "One Slogan, Many Methods_ Black Lives Matter Enters Politics," The New York Times - November 18, 2015

118 Ibid

Chapter 5:
119 "Job Openings and Labor Turnover Summary," U. S. Bureau of Labor Statistics, March 17, 2016

120 John F. Sargent Jr., "The U.S. Science and Engineering Workforce: Recent, Current, and Projected Employment, Wage and Unemployment," Congressional Research Service, February 19, 2014

121 Bob Costello and Rod Suarez, " Truck Driver Shortage Analysis 2015", American Trucking Association, October 2015

122 Virginia Harrison, "The world needs more pilots - 28,000 new jobs a year," CNN Money, July 21, 2015

123 Luke Stangel, "Facebook's 12 most fantastic employee perks," Silicone Valley Business Journal, April 8, 2013

124 Julie Bort, "LinkedIn is giving its employees 'unlimited' vacation plus 17 paid holidays," Business Insider, October 10, 2015

125 Spencer Soper, "Amazon Expands Medical Leave Benefits for employees," Bloomberg Business, November 2, 2015

126 Jim Puzzanghera, "Warren Buffett says tax hikes won't stop wealthy from investing," LA Times, November 26, 2012

127 Joel Andreas, "China - 'Smashing the iron rice bowl' -- expropriation of workers and capitalist transformation, International Journal of Socialist Renewal, October 2011 accessed October 15, 2015 http://links.org.au/node/2536

128 Christina Rogers, "U.S. Car Exports Top 2 Million," The Wall Street Journal, February 5, 2015

129 "Airbus U.S. plant cheaper than France, Germany, CEO tells paper," Reuters, September 12, 2015

130 Hiroko Tabuchi, "Chinese Textile Mills are now hiring in places where cotton was King," The New York Times, August 2, 2015

131 Mark Whitehouse, "Still Missing: 2.9 Million U.S. Workers," Bloomberg View, viewed on September 27, 2015

132 "Labor Force Statistics from the Current Population Survey – 2005-2015," U. S. Bureau of Labor Statistics, accessed November 8, 2015 http://data.bls.gov/pdq/SurveyOutputServlet

133 Mark Whitehouse, "Still Missing: 2.9 Million U.S. Workers," Bloomberg View, viewed on September 27, 2015

134 "What's Postmates," - Delivery in one hour, Accessed March 17, 2016 https://postmates.com/about

135 Charlie Rose, "Amazon's Bezos looking to the future," 60 Minutes, December 1, 2013

136 "The Sharing Economy," PWC.com, Consumer Intelligence Series (2015), 9

137 "Freelancing America: A National Survey of the New Workforce" Study Commissioned by Freelancers Union and Elance-oDesk, March 2014, accessed October 12, 2015 http://fu-web-storage-prod.s3.amazonaws.com/content/filer_public/c2/06/c2065a8a-7f00-46db-915a-2122965df7d9/fu_freelancinginamericareport_v3-rgb.pdf

138 Bob Sullivan, "For in-demand workers, it is not just about the money," CNBC October 14, 2014

139 Dareen Dahl, "Wonolo Hits Its Stride: How a Startup is Solving the On-Demand Staffing Dilemma, One Job at a Time," Coca-Cola Journey, Innovation, April 10, 2015, accessed October 14, 2015 http://www.coca-colacompany.com/innovation/wonolo-hits-its-stride-how-a-startup-is-solving-the-on-demand-staffing-dilemma-one-job-at-a-time

140 "Wonolo's Vision"   accessed October 14, 2015 http://wonolo.com/about-wonolo/

Chapter 6:

141 Erin Stepp, "Annual Cost to Own and Operate a Vehicle Falls to $8,698, Finds AAA," AAA NewsRoom, April 28, 2015

142 "Consumer Expenditures (Annual) News Release," U. S. Bureau of Labor Statistics, Economic New Release, September 3, 2015

143 "How Women Influence Healthcare Decisions," Healthgrades, June 2012

144 "Employed persons by detailed occupation, sex, race, and Hispanic or Latino ethnicity," Household Data Annual Averages, Table 11 - U. S. Bureau of Labor Statistics, 2014, Modified February 12, 2015, accessed November 16, 2015 http://www.bls.gov/cps/cpsaat11.htm

145 A.H.E.M. Mass and Y.E.A Appelman, "Gender differences in coronary heart diseases," Netherlands Heart Journal, Volume 18, Number 12, December 2010, 598

146 "Sex Specific Medical Research- Why Women's Health Can't Wait," - Brigham and Women's Hospital, Executive Summary, 2014

147 "What health issues or conditions affect women differently than men," National Health Institute, Last reviewed November 30, 2012, accessed November 15, 2015 https://www.nichd.nih.gov/health/topics/women-shealth/conditioninfo/Pages/howconditions.aspx

148 "Sex Specific Medical Research- Why Women's Health Can't Wait," - Brigham and Women's Hospital (2014), 7

149 "Percentage of women leading medical research studies rises, but still lags behind men," - Massachusetts General Hospital, News Release, July 19, 2006

150 "National Health Expenditures; Aggregate, Annual Percent Change, Percent Distribution and Per Capita Amounts, by Type of Expenditure: Selected Calendar Years 1960-2013," Centers for Medicare and Medicaid Service (CMS), NHE Tables, Historical Data, Table 2

151 "Former healthcare CEO argues America's medical system rewards bad outcomes," PBS NewsHour, Jan 17, 2014

152 Mark J. Perry, "Government Funding Increases Healthcare Costs," AEIdeas, American Enterprise Institute, January 6, 2010

153 Ibid

154 Jen Wieczner, "Should health insurance be like car insurance," Market Watch - January 23, 2013

155 Ceci Connolly and David Hilzenrath, "Obama Endorses Health Industry's Goal to Rein In Costs," The Washington Post, May 12, 2009

156 "Employer Health Benefits – 2015 Summary of Findings, Exhibit G," Kaiser Family Foundation 2015

157 "Stop the price-gouging at the for-profit Hospital Corporation of America," Joe Weinzettle at Change.com, accessed on October 31, 2015 https://www.change.org/p/federal-trade-commission-stop-the-price-gouging-at-the-for-profit-hospital-corporation-of-america

158 David U. Himmelstein, MD, Deborah Thorne, PhD, Elizabeth Warren, JD, Steffie Woolhandler, MD, MPH "Medical Bankruptcy in the United States, 2007 - Results of a National Study," Table 2 - The American Journal of Medicine, Vol 122, No. 8 - August 2009

159 "Up to 40 percent of annual deaths from each of five leading US causes are preventable," Centers for Disease Control and Prevention, Press Release May 1, 2014

160 Jessica Davis, "Healthcare in 2016: consumer-driven," HealthcareITNews, November 20, 2015

Chapter 7:

161 Lindsey Dunn, "Women's Executive Leadership Still Lags, And It Matters More in Healthcare Than Other Industries," Hospital Review, April 24, 2014

162 Mike Ross, "Running a hospital is a woman's job" The Boston Globe, June 21, 2015

163 "Former healthcare CEO argues America's medical system rewards bad outcomes," PBS NewsHour, Jan 17, 2014

164 Rosemary Gibson and Janardan Singh, "The Battle over healthcare" (Rowman and Littlefield 2012), 67

165 Rhonda Collins, "Bring Nurses Back to the Bedside," For the Record, Vol 27 No.9, September 2015, 10

166 Eugene Kim, "Check Out This Futuristic Hospital With Roaming Robots Funded By Salesforce CEO Marc Benioff," Business Insider - January 29, 2015

167 Marilynn Marchione, "Studies show promise for a new generation of wireless pacemakers implanted without surgeries," AP-USA News - November 9, 2015

168 Dave Fornell, "The Uberization of Healthcare," Diagnostic and Interventional Cardiology (DAIC) Magazine, September 4, 2015

169 Brian Honigman, "7 Biggest Innovations in Healthcare Technology in 2014" ReferralMD.com, November 2013

170 John T. James, A New, Evidence-based Estimate of Patient Harms Associated with Hospital Care," Journal of Patient Safety (September 2013), Volume 9, Issue 3, 122-128

171 Viktor Mayer-*Schönberger* and Kenneth Cukier, Big Data: A Revolution that will Transform How We Live, Work and Think (Houghton, Mifflin Harcourt 2013), 1-3

172 Jeremy Ginsberg Matthew H. Mohebbi Rajan S. Patel Lynnette Brammer Mark S. Smolinski Larry Brilliant, "Detecting influenza epidemics using search engine query data," Google Archives, originally published in Nature Vol 457, February 2009

173 Adam Rubenfire, "Hospitals use big-data platform to improve care," Modern Healthcare, August 22, 2015

174 "Carnegie Mellon Researchers Hack Off-the-Shelf 3-D Printer Toward Rebuilding the Heart," Carnegie Mellon College of Engineering, accessed October 27, 2015 http://engineering.cmu.edu/media/feature/2015/10_23_feinberg_paper.html

175 Andrew Zaleski, "Bioprinting: The new frontier in medicine that's not science fiction," CNBC, November 2, 2015

176 Michael Craig, "Could 3-D Printed Organs Be The Future of Medicine," Forbes, March 31, 2014

177 Chad Terhune, "Many hospitals, doctors offer cash discount for medical bills," LA Times, May 27, 2012

Chapter 8:

178 "Medical Profession: The Study of Women and Men," Center for Research on Gender in Professions, University of California, Davis, accessed on November 15, 2015 http://crgp.ucsd.edu/documents/GenderinMedicalProfessionsCaseStudy.pdf

179 Dave Fornell, "The Uberization of Healthcare," Diagnostic and Interventional Cardiology (DAIC) Magazine, September 4, 2015

180 Nicki Howell, "The Doctor's Office of 2024_ 4 Predictions for the Future," SoftwareAdvice.com, A Gartner Company, The Profitable Practice Blog, May 5, 2014

181 Pauline Chen, "Doctors and Patients, Lost in Paperwork," The New York Times, April 8, 2010

182 Lisa Zamosky, "Doctors are shifting their business models," LA Times, August 31, 2014

183 "See a doctor anytime, from anywhere," VirtualHealthNow.com, accessed on November 4, 2015 http://www.myalliancehealth.com/virtualhealthnow/

184 Felice J. Freyer, "It costs you $43 every time you wait for the doctor," The Boston Globe, October 5, 2015

185 Judy Packer-Tursman, "The Defensive Medicine Balancing Act," Medical Economics, January 9, 2015

186 "Quantifying the cost of Defensive Medicine – Summary of Findings," - Jackson Healthcare, February 2010

Chapter 9:

187 "Diagnostic & Medical Laboratories in the US – Market Research Report," IBISWorld, October 2015

188 Magellan Diagnostics, accessed June 2, 2016

189 "Nanobiosym's Gene-RADAR® Unveiled as the 'Next Frontier' in Healthcare Technology at the Clinton Global Initiative Annual Meeting," Business Wire, September 30, 2015

190 Andrew Pollack, "Jennifer Doudna, A Pioneer Who Helped Simplify Genome Editing", The New York Times, May 11, 2015

191 "Personalized Medicine Overview," US News content developed by Duke-Medicine, Last reviewed on January 20, 2011, accessed November 22, 2015 http://health.usnews.com/health-conditions/cancer/personalized-medicine/overview

192 "The Case for Personalized Medicine" Personalized Medicine Coalition, 4th edition 2014, 15

193 Christina Farr, "Why 3D Printing has 'Tremendous Potential' for Big Pharma," KQED, August 6, 2015

194 Ashlee Vance, "Illumina's New Low-Cost Genome Machine Will Change Healthcare Forever," Businessweek, January 15, 2014

195 Antonio Regalado, "Editas CEO Katrine Bosley Predicts First CRISPR Human Trial in 2017," MIT Technology Review, November 5, 2015

196 "Innovation in Drug Development Process Remains a Key Challenge," Tufts Center for the Study of Drug Development, October 20, 2015

197 John LaMattina, "The Debate Over The Cost To Develop A New Drug," Forbes, December 2, 2014

198 "FDA Basics - What are unapproved drugs and why are they on the market," FDA - accessed November 26, 2015 http://www.fda.gov/AboutFDA/Transparency/Basics/ucm213030.htm

199 Michael Hiltzik, "The little-known FDA program that's driving drug prices higher," LA Times, September 23, 2015

200 "Understanding Investigational Drugs and Off Label Use of Approved Drugs," U.S. Food and Drug Administration, accessed March 19, 2016 http://www.fda.gov/ForPatients/Other/OffLabel/default.htm

201 "Eliminating Pay-for-Delay Pharmaceutical Settlements Would Save Consumers $3.5 Billion annually," Federal Trade Commission, June 23, 2009

202 "Is it legal for me to personally import drugs?" Food and Drug Administration, accessed November 26, 2015 http://www.fda.gov/AboutFDA/Transparency/Basics/ucm194904.htm

203 Rosemary Gibson and Janardan Singh, "The Battle over healthcare" (Rowman and Littlefield 2012), 22

204 Bradley J. Fikes, "Costly Turing drug gets competition," San Diego Union Tribune, October 22, 2015

205 "AMA Calls for Ban on Direct to Consumer Advertising of Prescription Drugs and Medical Devices," American Medical Association - Press Release - November 17, 2015

206 C. Lee Ventola, "Direct-to-Consumer Pharmaceutical Advertising - Therapeutic or Toxic," Pharmacy &Therapeutics, Vol 36 No.10, October 2011

207 AMA Calls for Ban on Direct to Consumer Advertising of Prescription Drugs and Medical Devices - American Medical Association - Press Release - November 17, 2015

Chapter 10:

208 "How Women Influence Healthcare Decisions," Healthgrades, June 2012

209 Carolyn Buck Luce and Julia Taylor Kennedy, "The Healthcare Industry Needs to Start Taking Women Seriously," Harvard Business Review, May 28, 2015

210 "The Healthcare Imperative: Lowering Costs and Improving Outcomes: Workshop Series Summary, Excess Administrative Costs," National Academy of Sciences (National Academies Press 2010), Chapter 4, 7, http://www.ncbi.nlm.nih.gov/books/NBK53942/

211 Viktor Mayer-*Schönberger* and Kenneth Cukier, Big Data: A Revolution that will Transform How We Live, Work and Think (Houghton, Mifflin Harcourt 2013), 1-3

212 "Medicare Provider Utilization and Payment Data," Centers for Medicare and Medicaid Services, April 30, 2015

213 Chad Terhune, "Hospitals cut some surgery prices after CalPERS caps reimbursement," LA Times - June 23, 2013

214 "The ACA is Working for Women," U.S. Department of Health and Human Services, accessed on November 29, 2015 http://www.hhs.gov/healthcare/facts-and-features/fact-sheets/aca-working-women/index.html

215 "Obama and Clinton at CGI Obamacare forum (text, video)," provided by the White House, POLITICO Staff - September 24, 2013

216 Amy Goldstein, "White House projects marginal ACA enrollment growth in 2016," The Washington Post, October 15, 2015

217 Laura Heller, "Obamacare Is Turning Walmart Workers Into Temps" Forbes, June 14, 2013

218 Tami Luhby, "Employers play Obamacare blame game," CNNMoney, August 29, 2013

219 Susan Levine and Amy Goldstein, "Two more Obamacare health insurance plans collapse," Washington Post, October 16, 2015

220 Chad Terhune, "UnitedHealth may dump Obamacare plans, putting California expansion in doubt," LA Times, November 20, 2015

Chapter 11:

221 "Status of the Social Security and Medicare Programs – A summary of the 2015 Annual Report," Social Security Administration,3, accessed November 27, 2015 https://www.ssa.gov/oact/trsum

222 "Better, Smarter, Healthier: In historic announcement, HHS sets clear goals and timeline for shifting Medicare reimbursements from volume to value," Press Release, January 26, 2015

223 Alina Salganicoff, "Women and Medicare: An unfinished Agenda," Journal of American Society of Aging Blog, June 4, 2015

224 " NHE Fact Sheet – Historical NHE: 2013," Centers for Medicare & Medicaid Services, last modified July 28, 2015, accessed November 28, 2015 https://www.cms.gov/research-statistics-data-and-systems/statistics-trends-and-reports/nationalhealthexpenddata/nhe-fact-sheet.html

225 "Using lower-cost bevacizumab to treat wet AMD and diabetic macular edema could save Medicare billions," University of Michigan, Kellogg Eye Center, June 2, 2014 accessed November 18, 2015 http://www.kellogg.umich.edu/news/14/choosing-avastin-over-lucentis-could-save-medicare-billions.html

226 Laura Stokowski, "The Compounding Controversy," Medscape, July 3, 2013

227 "Generic Drug Savings in the U.S., Seventh Annual Edition: 2015," Generic Pharmaceutical Association (GPhA), 5

228 "Candidate Barack Obama's primary campaign Ad 2008" accessed April 10, 2016 https://www.youtube.com/watch?feature=player_embedded&v=NCRO0g9CfAw

229 Marc-André Gagnon, PhD. and Sidney Wolfe, MD., "Mirror, Mirror on the Wall: Medicare Part D pays needlessly high brand name drug prices compared with other OECD countries and with U.S. government programs," Public Citizen," Carlton University, Policy Brief, Fig 5, July 23, 2015

230 Kathleen M. King, "Medicare Fraud – Progress Made, but More Action Needed to Address Medicare Fraud, Waste and Abuse," GAO Testimony Before the Subcommittee on Health, Committee on Ways and Means, House of Representatives, April 30, 2014

231 Chad Terhune, Noam Levey and Doug Smith, "Release of Medicare doctor payments shows some huge payouts," LA Times, April 8, 2014

Chapter 12:

232 "Fast Facts – Teacher Trends," National Center for Education Statistics, accessed January 12, 2016 http://nces.ed.gov/fastfacts/display.asp?id=28

233 "Characteristics of Public and Private Elementary and Secondary School Principals in the United States: Results from the 2011-12 Schools and Staffing Survey, First Look," NCES - 2013-313, U.S. Department of Education

234 Carla Rivera, "Two women named to lead Cal State campuses Chico and Channel Islands," LA Times, March 9, 2016

235 Cathy Sandeen, "If a female president is good for the Ivy League, why not for the rest of us," The Conversation, August 6, 2015

236 Regina L. Garza Mitchell and Pamela L. Eddy, "Moving UP or Moving On- A Gender Perspective on Mid-Level University Leaders," Journal of Higher Education Management, Volume 30 (2015) 65-81

237 "Fact Sheet: Celebrating Seventy-Five Years of International Exchanges," The UNITED STATES Department of State's Bureau of Educational and Cultural Affairs (ECA), accessed October 25, 2015 http://eca.state.gov/files/bureau/factsheet_eca.pdf

238 Tom Segal, "Rethinking the Learning Experience: Part IV," Huff Post, August 9, 2012

239 Salman Khan, *The One World Schoolhouse* (Hachette 2012), 179

240 Steve Denning, "The Single Best Idea for Reforming K-12 Education," Forbes, September 1, 2011

241 Salman Khan, *The One World Schoolhouse* (Hachette 2012), 52

242 Ibid, 69

243 Lael Brainard, "Coming of Age in the Great Recession," Board of Governor of the Federal Reserve System, Remarks At the "Economic Mobility: Research and Ideas on Strengthening Families, Communities, and the Economy" Ninth Biennial Federal Reserve System Community Development Research Conference, Washington, D.C. - April 2, 2015

244 Newt Gingrich, *To Renew America* (Harpercollins 1996), 142

245 David O. Lucca, Taylor Nadauld and Karen Shen, "Credit Supply and the Rise in College Tuition: Evidence from the Expansion in Federal

Student Aid Programs," Federal Reserve of New York, Staff Report 733, July 2015, 26

246 "Expenditures of educational institutions, by level and control of institution - Selected years, 1899-1900, Table 106.20, Digest 2014," National Center for Education Statistics, accessed April 10, 2016 https://nces.ed.gov/programs/digest/d14/tables/dt14_106.20.asp

247 "Enrollment in elementary, secondary, and degree-granting postsecondary institutions by level and control of institution, 1869 - 2024, Table 105.30," National Center for Education Statistics, accessed April 10, 2016 https://nces.ed.gov/programs/digest/d14/tables/dt14_105.30.asp

248 "Public Education Finances: 2013, Economic Reimbursable Surveys Division Reports," Education Finance Branch, U.S. Census, Fig. 1, G13-ASPEF, June 2015

249 Evan Thomas, "Why We Must Fire Bad Teachers" Newsweek, March 5, 2010

Chapter 13:

250 Salman Khan, *The One World Schoolhouse* (Hachette 2012), 76

251 Matt Miller, *The tyranny of dead ideas:* by, Revolutionary Thinking for a New Age of Prosperity, (St. Martin's Griffin 2010), 108

252 "Total and current expenditures per pupil in public elementary and secondary schools: Selected years, 1919-20 through 2011-12, Table 236.55," National Center for Education Statistics, Digest 2014, accessed January 1, 2016 http://nces.ed.gov/programs/digest/d14/tables/dt14_236.55.asp

253 Paul E. Peterson, Ludger Woessmann, Eric A. Hanushek, Carlos X. Lastra-Anadón, "Globally Challenged - Are U. S. Students Ready to Compete?," Harvard Kennedy School, PEPG Report No:11-03, August 2011, 19

254 "Fact Sheet 2014 - School Administrators: An Occupational Overview," AFL-CIO, October 2014

255 Theresa Watanabe and Howard Blume, "Michelle King is new superintendent for Los Angeles Unified School District," LA Times, January 11, 2016

256 Abby Jackson, "Mark Zuckerberg's $100 million donation to Newark public schools failed miserably—here's where it went wrong," Business Insider, September 25, 2015

257 Howard Blume, "New report finds ongoing iPad and technology problems at L.A. Unified" LA Times, September 2, 2015

258 Howard Blume, "Charter school expansion could reshape L.A. Unified, officials say," LA Times, November 12, 2015

259 Alex Konrad, "With $100 Million From Silicon Valley Elite, AltSchool Takes New Approach To Classroom Learning," Forbes, May 4, 2015

260 Anya Kamenetz, "A For-Profit School Startup Where Kids Are Beta Testers," NPR, May 4, 2015

261 "Khan Academy – About Us," accessed April 10, 2016 https://www.khanacademy.org/about

262 "8.2 million unique visitors in September 2015 - Khan Academy" accessed October 31, 2015 https://siteanalytics.compete.com/khanacademy.org/#.VjV-PNK6CUl

263 Kim Lachance Shandrow, "Khan Academy Founder - No, You're Not Dumb. Anyone Can Learn Anything," Entrepreneur, October 30, 2015

264 "Gates Foundation - College Ready Strategy," accessed October 31, 2015 http://collegeready.gatesfoundation.org/about/

265 Arik Hesseldahl, "Why Oracle Is Building a Public High School on Its Campus," <re/code>, October 27, 2015

266 Izzy Best, "Etsy for teachers - TpT becomes hub for education materials," CNBC, October 11, 2015

267 Natasha Singer, "Google Virtual-Reality System Aims to Enliven Education," New York Times, September 28, 2015

268 Randi Weingarten and Stanley Litow, "Career and Technical Education Programs Provide Path to Middle-Class Jobs," by US News, Opinion, December 18, 2015

Chapter 14:

269 Salman Khan, *The One World Schoolhouse* (Hachette 2012), 233

270 "Expenditures of public degree-granting postsecondary institutions – 2005-2012, Table 334.10," National Center for Education Statistics,

Digest 2013, accessed January 13, 2016 https://nces.ed.gov/programs/digest/d13/tables/dt13_334.10.asp

271 "Benchmarking Women's Leadership in the United States," University of Denver, August 2013, 13

272 "Leadership Institutes Building Pipelines of Women, Minorities," Diverse - August 3, 2015

273 "Benchmarking Women's Leadership in the United States," University of Denver, August 2013, 20

274 Ellie Bothwell, "World's top 10 universities led by women," Times Higher Education (THE), June 19, 2015

275 Jason Song, "Higher Learning Professors question traditional four-year residential college model," LA Times, Feb 25, 2015

276 "For 1st time, MIT's free online classes can lead to degree," The Associated Press, October 7, 2015

277 Lauren Camera, "Colleges Slash Tuition to Eliminate Sticker Shock," by US News, September 17, 2015

278 "These 30 Colleges are Reversing the Rise in Tuition," AffordableSchools. net, accessed January 13, 2016 http://affordableschools.net/30-colleges-reversing-rise-tuition/

279 "Fast Facts - Graduation Rates," National Center for Education Statistics, November 7, 2015 https://nces.ed.gov/fastfacts/display.asp?id=40

280 Gregory Ferenstein, "Khan Academy founder has two big ideas for over-hauling higher education in the sciences," Venturebeat.com, December 12, 2014 (unable to reverse linked to 282 in content)

281 Gregory Ferenstein, "Why Google doesn't care about college degrees, in 5 quotes," Venturebeat.com, April 25,2014

Chapter 15:

282 Blake Ellis, "40 million Americans now have student loan debt," CNN, September 10, 2014

283 "15 Economic Facts about Millennials, Fact 7," The Council of Economic Advisers, October 2014, 18

284 "The Precarious State of Family Balance Sheet," Pew Charitable Trusts Report, January 2015, 9

285 Lawrence Lewitinn, "Why the housing market could soon get positive," Yahoo Finance, accessed September 30, 2015 http://finance.yahoo.com/news/why-the-housing-market-could-soon-get-positive-news-172927524.html

286 Tim Ahern, "What Will Housing Demand Look Like in the Future-What You Need to Know," Housing Industry Forum, Fannie Mae, June 4, 2015

287 Michael J. Silverstein and Kate Sayre, "The Female Economy," Harvard Business Review, September 2009, 3

288 Christina Gritmon, "More single women are 1st-time homeowners," USA Today-The (Westchester County, N.Y.) Journal News, August 23, 2015

289 "For Women, Student Loan Debt Is an Even Bigger Crisis," American Association of University Women, July 8, 2014

290 Jillian Berman, "If you live in this state, you could have your student loans forgiven," MarketWatch.com, January 5, 2016

291 Mary Umberger, "Housing industry pins its hopes on millennials," LA Times, Jan 18, 2015

292 Susan Dynarski, "We're Frighteningly in the Dark About Student Debt," The New York Times, March 20, 2015

293 David Jesse, "Government books $41.3 billion in student loan profits," Detroit Free Press - USA Today, November 25, 2013

294 Mandi Woodruff, "What SoFi's $1 billion news means for the student loan refi business," Yahoo Finance, October 1, 2015

295 "Financial wellness at work – A review of promising practices and policies," Consumer Financial Protection Bureau, August 2014, 9

296 Robert Powell, "More companies offer financial wellness programs," USA Today, September 27, 2014

297 Bobbi Rebell, "The latest U.S. corporate perk - Student loan help," Reuters, September 22, 2015

298 "Income-Driven Plans Federal Student Loan," Department of Education, accessed October 18,, 2015 https://studentaid.ed.gov/sa/repay-loans/understand/plans/income-driven

299 Jillian Berman, "What's really causing the student debt crisis, according to Sheila Bair," MarketWatch.com October 17, 2015

300 Mary Kate Cary, "Why the Government is to Blame for High College Costs," US News, November 23, 2011

301 Jenai Engelhard Humphreys, "What can't grads afford because of student loans- Kids for one," Boston Globe, May 22, 2015

302 "Is Student Debt Prolonging the Recession?," Equal Justice Works, U.S. News, March 7, 2012

Chapter 16:

303 "Tax Freedom Day 2013: Data Tables, Historical Tax Freedom Day, 1900-2013," Tax Foundation, accessed January 14, 2016 http://taxfoundation.org/article/tax-freedom-day-2013-data-tables

304 Jared Meyer, "Working for the Tax Man," The Manhattan Institute, April 16, 2015

305 Michael Tasselmyer, "A Complex Problem – The Compliance burden of tax code," National Taxpayers Union Foundation, Policy Paper No. 176, April 8, 2015, 5

306 "April 15th_ The joy of tax – A futile plea for simplicity," The Economist, April 8, 2010

307 Michael Tasselmyer, "A Complex Problem – The Compliance burden of tax code," National Taxpayers Union Foundation, Policy Paper No. 176, April 8, 2015, 3

308 Stephen Ohlemacher, "Taxes Take Up More Than 6 Billion Hours Of America's Time Every Year," Associated Press, January 9, 2013

309 Michael Tasselmyer, "A Complex Problem – The Compliance burden of tax code," National Taxpayers Union Foundation, Policy Paper No. 176, April 8, 2015, 1

310 Arthur Lafffer, Wayne H. Winegarden and John Childs, "The Economic Burden Caused by Tax Code Complexity," Executive Summary, The Laffer Center, April 2011

311 Fareed Zakaria, "Global Public Square - Interview with Bill Gates," CNN, May 17, 2015

312 "Fiscal 2015 Budget: Special Topics," Office of Management and Budget, Table 15-1, 245

313 Chris Edwards, "Downsizing the Federal Government - Fiscal Federalism," CATO Institute, June 2013, 2

Chapter 17:

314 "Warren Buffett and His Secretary on Their Tax Rates," ABC News, January 25, 2012

315 David Lazarus, "Average taxes on wireless bills in California reach a record 18%," LA Times, November 24, 2015

316 "Gasoline Taxes," American Petroleum Institute, accessed April 11, 2016 http://www.api.org/oil-and-natural-gas/consumer-information/motor-fuel-taxes/gasoline-tax

317 "Contribution and Benefit Base," Social Security Administration, accessed January 2, 2016 https://www.ssa.gov/oact/COLA/cbb.html#Series

318 Fig 17.1 - Tax on income from $37405 to $90,750 ($5156) Plus 25% over $37450 ($3138) plus 7.65% Payroll Tax ($3825) = Total for working income is $12118.75; non-working income is 20% flat = $10,000 and corporate tax is 15% flat = $$7.500

319 "Who Pays: A Distributional Analysis of the Tax Systems in All Fifty States, Fifth Edition," Institute on Taxation and Economic Policy, Executive Summary, January 2015, 1

320 "Tax Policy and the Economy - Volume 26, Table 2," - National Bureau of Economic Research, July 2012, 99

321 "All Returns - Selected Income and Tax Items in Current and Constant 1990 Dollars, Tax Years 1990 - 2013 - Table A," IRS, accessed April 11, 2016 https://www.ssa.gov/oact/COLA/cbb.html#Series

322 "Estimated Average Taxpayer Burden for Individuals by Activity," IRS, accessed October 21, 2015 https://www.irs.gov/instructions/i1040ez/ar03.html

323 "All Returns - Selected Income and Tax Items in Current and Constant 1990 Dollars, Tax Years 1990 - 2013 - Table A," IRS, accessed April 11, 2016 https://www.ssa.gov/oact/COLA/cbb.html#Series

324 Evan Halper, "State Will Do Taxes for Some," LA Times, February 20, 2005

325 "The Distribution of Household Income and Federal Taxes, 2011, Table#4, Fig. 4," Congressional Budget Office, November 2014, 16

Chapter 18:

326 "Table 2.1 Receipts by Source - 1934- 2017," Office of Management and Budget, accessed October 22, 2015 http://www.whitehouse.gov/omb/budget/Historicals

327 David S. Logan, "Fiscal Fact: Three Differences Between Tax and Book Accounting that Legislators Need to Know," Tax Foundation, No. 277, July 27, 2011

328 "Taxpayer Compliance Costs for Corporations and partnerships: A New Look," IRS, accessed October 24, 2015, 6 https://www.irs.gov/pub/irs-soi/12rescontaxpaycompliance.pdf

329 "Corporate Tax Expenditure – Information on Estimated Revenue Losses and Related Federal Spending Programs," General Accounting Office (GAO), GAO-13-339, March 2013, 10

330 "CHARTBOOK – Offshore Corporate Taxes, Corporate Profits and the Competitiveness of the U.S. Tax System," Americans for Tax Fairness - September 2015

331 Dominic Rushe, "Apple saves $9bn with bond manoeuvre," The Guardian, May 1, 2013

332 Zachary R. Mider, "Here is How CEOs Flee Taxes While Staying in U.S.," Bloomberg Business, May 5, 2014

333 Robert Cyran, "Pfizer's Huge Deal Could Work, but With Adverse Effects," The New York Times, November 23, 2015

334 "The Taxation of Carried Interest – Statement of Peter Orszag, Director Congressional Budget Office Testimony before the Committee on Finance United States Senate, July 11, 2007," 3,2

335 Donald L. Bartlett and James B. Steele, *The Betrayal of American Dream*, (Public Affairs October 2013) P. 22

336 "Leveling the playing field - Curbing tax havens and removing tax incentives from shifting jobs overseas," The White House, Office of the Press Secretary, March 4, 2009

337 Shobhana Chandra, "Economy in U.S. Picked Up on Consumer Spending, Construction," Bloomberg Business, September 25, 2015

338 Karen Brettell, David Gaffen and David Rohde, "Corporate America's buyback binge feeds investors, starves innovation," *Reuters,* November 16. 2015

Chapter 19:

339 "Congressional Performance, Hate Congress, Love Your Congressman? Not Quite," Rasmussen Reports™ September 10, 2015

340 "Microsoft Employee Town Hall with Erskine Bowles," December 7, 2012 accessed September 29, 2014 http://www.microsoft.com/en-us/news/speeches/2012/12-07fixthedebt.aspx

341 Andy Serwer, "BlackRock's Fink says our biggest risk isn't China or interest rates_ It's us," Yahoo Finance, May 27, 2015

342 Munir Moon, The *Beltway Beast* (MGN Books 2014), 3

343 "Washington is the biggest threat to US security Robert Gates," Robert Gates," –Global Public Square with Fareed Zakaria, CNN, January 19, 2014

344 Mark Leibovich, This Town: Two Parties and a Funeral-Plus, Plenty of Valet Parking!-in America's Gilded Capital, (Blue Rider Press, April 29, 2014), Pg.12

345 Carmen DeNavas-Walt and Bernadette D. Proctor, "Income and Poverty in US – 2013 – Current Population Report," Fig. 4, U.S. Census – September 2014, 12

346 Walter Alarkon, "Self-Imposed Republican moratorium leads to drop in 2011 earmark spending," The Hill, August 1, 2010

347 "Fact Sheets - Earmarks and Earmarking_ Frequently Asked Questions," TaxpayersforCommonSense,accessedFebruary26,2013http://www.taxpayer.net/library/article/earmarks-and-earmarking-frequently-asked-questions#1

348 Jonathan Allen, "Members of Congress - This job sucks," POLITICO, March 7, 2012

349 Ronald D. Utt, Ph.D., "The Bridge to Nowhere_ A National Embarrassment," Heritage Foundation, October 20, 2005

350 Abbie Boudreau and Scott Bronstein, "The bridge failed, but the 'Road to Nowhere' was built," CNN, September 24, 2008

351 Richard Lardner, "Army says no to more tanks, but Congress insists," Associated Press - Yahoo! News April 28, 2013

352 Andrew Prokop, "Ohio voters just took a big step to fight gerrymandering," Vox, November 4, 2015

353 Richard Wolf and Gregory Korte, "Supreme Court strikes blow against gerrymandering," USA Today, June 29, 2015

354 Alex Tribou and Adam Pearce, "Courts Are Shaking Up House Elections in 2016," Bloomberg Politics, December 3, 2015

355 Mark Thompson, "The True Cost of the Afghanistan War May Surprise You," TIME, January 1, 2015

356 "Final State of the Union Address, President Obama" January 13, 2016, The White House

357 Stephen Vizinczey, "Afghanistan is an unwinnable war, and our leaders know it," Telegraph, August 2, 2010

358 Niall Ferguson, "The Kissinger Diaries: What He Really Thought About Vietnam," POLITICO Magazine, October 10, 2015

359 Marilyn B. Young, "Vietnam's Second Front: Domestic Politics, the Republican Party, and the War by Andrew L. Johns (review)," Journal of Cold War Studies, Vol. 14, Number 4, Fall 2012, 231-231

360 "Lyndon B. Johnson - Foreign Affairs," Miller Center, University of Virginia, accessed October 17, 2015 http://millercenter.org/president/biography/lbjohnson-foreign-affairs

361 Lydia Wheeler, "States rise up against Washington," TheHill, February 10, 2015

362 Patrick McGreevy, "Jerry Brown OKs automatic voter registration through DMV," LA Times, October 11, 2015

363 Wendy Weiser and Erik Opsal, "The State of Voting in 2014," Brennan Center for Justice, New York University School of Law, June 17, 2014

364 Jessica M. Vaughan "Number of Sanctuaries and Criminal Releases Still Growing - 340 Sanctuaries release 9,295 criminals," Center for Immigration Studies, October 2015

365 Lydia Wheeler, "States rise up against Washington," TheHill, February 10, 2015

366 John M. Glionna, "Bills seek to turn U.S. land over to 11 states - group raises questions," LA Times, August 11, 2015

367 Carl Hulse and David M Herszenhorn, "After Boehner, House Hard-Liners Aim to Weaken Speaker's Clout," The New York Times, October 10, 2015

368 "The Medicare Catastrophic Coverage Act of 1988 Staff Working Paper," Congressional Budget Office, October 1988, accessed September 30, 2015 https://www.cbo.gov/sites/default/files/100th-congress-1987-1988/reports/88doc14_0.pdf

Chapter 20

369 Julianna Goldman, "Donald Trump's Cabinet richest in U.S. history, historian say", CBS News, December 20, 2016

370 James Rickards, "Repeal of Glass-Steagall Caused the Financial Crisis," US News -Opinion, August 27, 2012

371 Alexander Burns and Jonathan Martin, "Republican Leaders Map a Strategy to Derail Donald Trump," The New York Times, March 19, 2016

372 Drew DeSilver, " "What is the House Freedom Caucus, and who's in it," Pew Research Center, October 20, 2015

373 Charles Wheelan, *The Centrist Manifesto* (W.W. Norton & Company 2013), 72

374 Munir Moon, The *Beltway Beast* (MGN Books 2014), 12

375 "Signatures needed for independent candidates to qualify for U.S. Congress" Ballotpedia.org, accessed March 23, 2016 https://ballotpedia.org/Signatures_needed_for_independent_candidates_to_qualify_for_U.S._House_of_Representatives_elections,_2014

376 Jeffrey Jones, "In U.S., New Record 43% are Political Independents," Gallup, January 7, 2015

377 Jennifer E. Manning, "Membership of the 114th Congress – A Profile," Congressional Research Service, January 22, 2015

378 "List of current United States Senators by age," Wikipedia. org, accessed April 11, 2016 https://en.wikipedia.org/wiki/List_of_current_United_States_Senators_by_age

379 Blake Ellis, "40 million Americans now have student loan debt," CNN, Sep. 10, 2014

380 Josh Mitchell, "Federal Student Lending Swells," Wall Street Journal, November 28, 2012

381 "Online Voter Registration," National Conference of State Legislatures, accessed March 28, 2016 http://www.ncsl.org/research/elections-and-campaigns/electronic-or-online-voter-registration.aspx

382 Dale Stephens, "The Rise of Political Entrepreneurship Uncollege. org blog, accessed November 25, 2015 http://blog.uncollege.org/the-rise-of-political-entrepreneurship

383 Edward Lopez, "What do we mean by Political Entrepreneur?" November 14, 2012 viewed April 11, 2016 http://politicalentrepreneurs.com/what-do-we-mean-by-political-entrepreneur/

Chapter 21

384 Scott Horsley, "American will always do the right things," NPR- October 28, 2013